# VAIROS

Data Driven

Strategic Planning

J. Tod Fetherling

Publisher

7011 Ellendale Drive

Brentwood, TN. 37027


ISBN: 979-8-9928531-1-7


December 1, 2025

# Acknowledgments

Book writing, regardless of the genre, is a journey. This expedition I embarked upon took time. It started nearly 10 years ago. In the beginning, I would work on it sporadically, for a few hours at most each year. It was only recently that I felt motivated to finish this book, post a strategy session that inspired me deeply.

I owe a massive debt of gratitude to my family. My wife, Mary, has been my steady companion through many late nights of writing and revision, offering patience and perspective when I needed them most. To Danny, Laura, Chase, and Carolyn—you have each encouraged me to pursue this dream and graciously given me the space and time to put these ideas on paper.

A few years ago, I moderated a panel entitled "From Big Data to Strategy" at the Summit of the Southeast for Tennessee HIMSS. The panel included Fred Trotter from DocGraph, Kevin Johnson, MD, MS from Vanderbilt University, and Arvind Kumar from Harvard University now with Eisner Amper. What we discovered was a delightful dialogue about what remains to be done: to translate data into actionable strategies.

The panelists revealed that while there were hopes about future revelations on standard models for converting massive datasets into actionable strategies, the capability itself remained largely undeveloped.

It was this revelation that gave birth to the Vairos framework—a systematic approach for bridging the gap between data and strategy.

One fundamental question emerged during this crucial discovery: What constitutes "big data"?

Dr. Kevin Johnson now with University of Pennsylvania provided the most relevant answer; "It depends." What constitutes big data for one organization might not do the same for another. Therefore, the definition is contextual and relative to your current data management capabilities.

Be it terabytes, petabytes, or exabytes- the real question here isn't about volume. Rather, the question is about readiness. Are we prepared to manage and extract strategic value from the data we have?

As Arvind Kumar noted, "The question we must really focus upon is of risk avoidance." You see, if we don't prepare to work systematically through data and information beginning now, we leave ourselves vulnerable to tremendous risk.

A special thank you to Beth Chase, the master doodler who helped me create the graphic of the model over lunch one day at Calypso Cafe.

To the broader community of thinkers, researchers, and practitioners who have contributed to the intersecting fields of this work—thank you for your dedication to advancing human knowledge.

Although I may not know all of you personally, yet your collective efforts have shaped the intellectual landscape that made this book possible.

Finally, I want to thank the readers who will engage with these ideas. The actual value of any book is not in its writing, but in its ability to inspire thought, dialogue, and action.

I hope the concepts discussed here will motivate you to see new connections, ask qualitative and relevant questions, and contribute to the ongoing conversation about how we can better understand and navigate our increasingly complex world.

Any errors or shortcomings in this work are entirely my responsibility. The wisdom and insights belong to the many who have generously shared their knowledge.

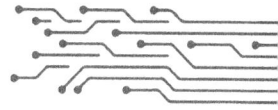

# Table of Content

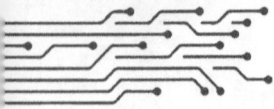

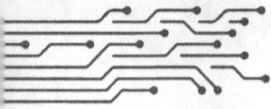

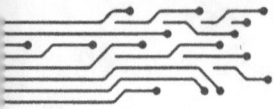

# Revisiting our Main Characters

Post the successful implementation of Omnimics at MegaHealth, Susan Myles gazed expansively out of her window. In a moment of flashback, she realized that nearly three months had gone by since the solution was rolled out.

The project was a resounding success undoubtedly, and Jim was thoroughly pleased with Susan and her efforts. As an additional testament to success, MegaHealth too was prospering like never before. However, it was now time to take things forward, because what good was an idea without execution?

Hence, Susan resolved to accelerate her efforts and combine forces with Jim to achieve his vision for the future.

The next morning, Susan woke up to a voice note from Jim in her inbox. Intrigued, she clicked on it immediately and heard Jim's voice, "Hey Susan, I remember you talked about Vairos when we were both working on that Omnimics roll out. Vairos, if I remember correctly from your description, is the future-proof approach to strategic planning. Along those lines, I feel I have found what we need to take things forward from here.

I'd like to introduce you to Megan Brooks. She's our Chairman of the Board, and we're beginning a new strategic planning process. I was wondering if the two of you could meet to discuss how we should approach a new data-driven strategy. Let me know what you think..."

Upon hearing this, Susan felt instantly excited at the prospect and called Jim back almost immediately. She said, "Count me in. I'm so excited to be able to work together, and would be available soon to meet with Megan."

When the meeting took place, Jim began by briefing Megan on the work he had done with Susan. "I am delighted to share that Susan's Omnimics approach has been somewhat groundbreaking. It has helped the organization truly understand the answer before questions were even asked. Her unique approach has enabled us to stay a step ahead of the data. The entire process is so good that we've almost become disciples of the concept and are quite versed in the explanation."

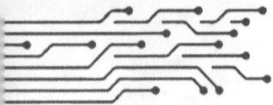

Jim went on to explain how the organization was now ready to grow under the influence of her approach. He continued to describe how Megan Brooks had approached Jim about creating a new strategic plan for the health system, including an Urgent Care Center, Primary Care Strategy, Inpatient expansion, Outpatient outposts, and Post-Acute Care.

Megan wanted to build a health system designed for the consumers rather than one that prioritized healthcare professionals over customer satisfaction. Considering what a challenging element Megan had targeted with this approach, there was no doubt the entire process would be quite rigorous and not without challenges.

Jim then shifted his focus to Susan and told her, "Megan had approached me earlier in the month and was interested in creating a new strategic plan for the health system."

He continued, "According to her health system plan, the primary focus will be on building an Urgent Care Center, Primary Care Strategy, Inpatient Expansion, and Outpatient outposts, and rethinking our Post-Acute Care unit."

He added, "Megan's health system plan focuses primarily on the consumer instead of the professionals in healthcare."

When Jim had concluded his explanation, Susan replied, "Jim, I have no doubt our path ahead is going to be rigorous and perhaps difficult. But you know what? I love a great challenge, and I sure am up for this one!"

Susan continued taking notes for an hour or so as Jim and Megan explained their concept and vision at length to her. At the end of it all, Megan asked, "So, Susan, I know it's a bit too soon, but do you have an idea for us to begin with yet?"

"You bet I do!" Susan replied, "Staying one step ahead is my core skill, ha-ha!" All three of them enjoyed Susan's little pun as she pulled up a diagram and said, "Alright, so this is basically a strategic approach I had in mind..., Jim, it basically integrates what you had already mastered with the Omnimics rollout. But what it does differently is that it walks you through a series of questions that are required for success."

"Is this the Vairos framework you were talking about earlier?" Jim asked. "Yes, this is the one!" said Susan. Jim would have rather launched straight into work than discuss more questions. Yet, he felt highly intrigued at the same time since he knew what Susan was capable of. So, he simply said, "Great, you have my undivided attention."

Meanwhile, Megan, who was observing the diagram closely, said, "It is simple and straightforward, but I am unfortunately not familiar with DIKW."

"Well, Megan, you see, the overall framework is intentionally designed to be clear in concept, but richer and more complex, in execution. The three key parts about this framework are: Assess, Decide, and Execute. It is imperative to understand each of these parts at length first, before you begin executing them", said Susan.

Susan noted Megan's peaked interested and continued, "Each of these parts connects with DIKW. I'll keep it simple for you. Data and Information constitute a significant part of the Assess phases; Knowledge is where you make the Decisions; and Wisdom is all about the execution. So, in essence, the whole DIKW thing is like a pyramid, but this one gets inverted when it comes to strategy."

Megan understood the basics but still had many questions that she had mentally noted down. However, time was running short for all three of them so there was no more time to discuss them at length presently.

Susan was the first to leave the boardroom. As the door closed behind her, Megan looked at Jim and said, "Why do we need a consultant when we have the framework?"

Jim, much too familiar with where the road leads, told her, "We don't! We don't need a consultant; what we need is a GUIDE, and that is exactly what Susan is. See, this is an expedition and this is exactly where Susan has the most value to offer, and I'm certain that all three of us will have a lot to learn in the weeks to come!"

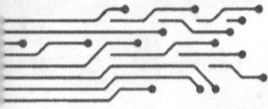

# Introduction to the Vairos Framework

## Strategic Planning in a Data-Driven World

### Introduction

The Vairos Planning System represents a fresh approach to strategic planning—one centered on the principle of Data Driven Strategy. In today's business environment, organizations are inundated with data yet find themselves with less time to focus on meaningful execution. Vairos offers a disciplined, methodical pathway to move any organization from its current state to a desired future, unlocking potential through strategic clarity and rigor.

# Vairos: Data Driven Strategic Planning

At its core, the Vairos Planning System rests on the understanding that effective strategy demands both analytical discipline and execution focus. The framework transforms the overwhelming complexity of modern data into a clear, actionable strategy through three interconnected phases:

- Victory in the right and opportune moment

- Securing victory with a well-crafted strategy

- Strategy as a process: assessing the situation, making decisions, executing the plan, and evaluating success

## The Vairos Framework

A few days later, all three of them met again in the boardroom at MegaHealth's HQ. During this meeting, the agenda was for Susan to explain the Vairos framework to Jim and Megan, and she had come prepared with presentations, data, charts, and whatnot.

As the meeting started, Susan pulled up a slide and started explaining that *"The Vairos Planning System translates a data continuum into a comprehensive strategic planning methodology. The framework operates through three major components: Assess, Decide, and Execute. Each component builds upon the previous, creating a systematic approach to data-driven strategy."*

While everyone seemed to be engrossed deep in thought during the meeting, Jim was the first to pose a question. *"What's with the name?" he said. "Is it some sort of abbreviation, or does it run deeper like Omnimics?"*

Susan replied, *"The name Vairos itself, embodies the philosophy: sealing one's victory in the right and opportune moment. Furthermore, it is imperative to understand that if you want victory, you need to have the right strategy first to achieve it.*

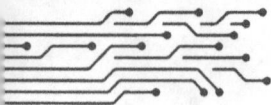

Secondly, you ought to note that strategy is just part of the bigger picture. The formula in entirety is to first have a strategy for victory in place and for a strategy you need to assess the situation, decide, execute a plan, and measure success against the objectives."

Jim assured Susan that he understood what she was getting at. However, he implied that he needed more details and time to evaluate them before he bought her idea.

Susan continued, "I know, and more detail is exactly what I have to offer today. The presentation I have here will guide you through each component of the Vairos system. But it doesn't end there. I have tools and frameworks that'll transform your organization and make it go from reactive to proactive.

In simple words, the process of transforming from a reactive to a proactive organization means having  the right data to help you make strategically aligned decisions which are ready for execution."

## The Strategic Data Continuum

The next slide Susan presented was a six-stage process layout. By this point it was evident that Susan had conducted detailed research on this model. The six stages she had outlined included:

- **Stage 1 - Discovery:** What do we have? What does it tell us? (Descriptive Statistics)

- **Stage 2 - Integration:** Market data and external systems (Diagnostic Statistics)

- **Stage 3 - Business Intelligence:** Internal business system reporting

- **Stage 4 - Predictive Analytics:** Perform, test, replicate (Agile Analytics)

- **Stage 5 - Execution:** Turning data into action (Prescriptive Analytics)

- **Stage 6 - Decision Sciences:** Capturing decisions based on available data and feeding back into Stage 1

**DATA DRIVEN STRATEGIC PLANNING**

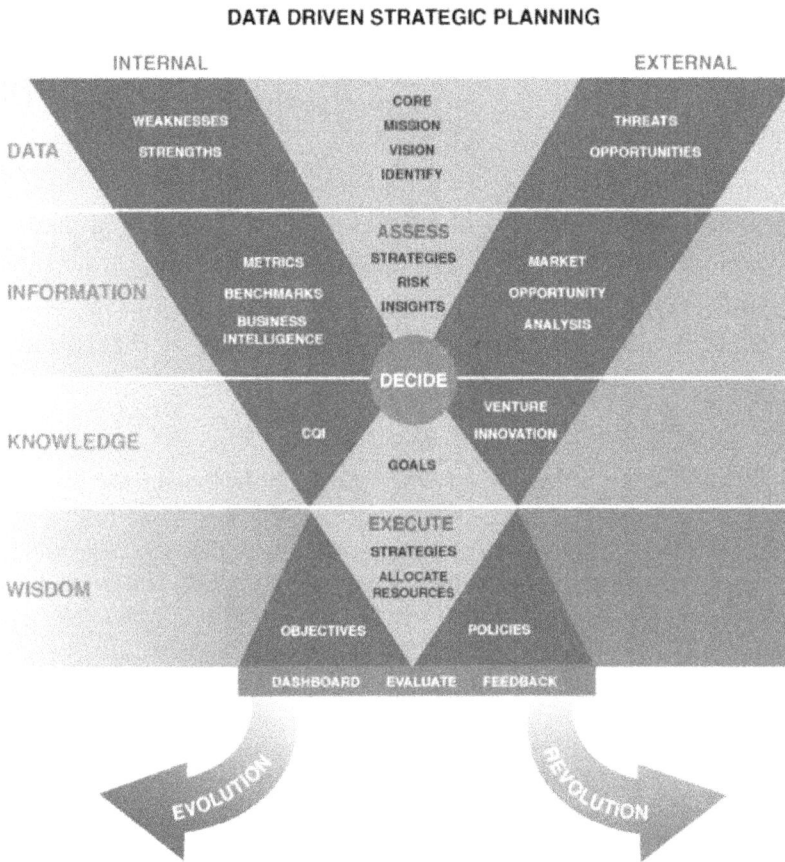

# The Three-Phase Architecture

## Phase 1: Assess

Glancing at the next slide, Susan commented that the Assessment Phase" is the foundation of all strategic work. When engaged in this phase, organizations conduct a comprehensive analysis of their current position.

"But what exactly do they analyze?" Jim wanted to know. "Glad you asked that" said Susan, "So in this phase, an organization examines both; its internal capabilities and external market conditions. This comprehensive analysis is highly important because it ensures that all the subsequent decisions of the organization are grounded in reality, not assumptions."

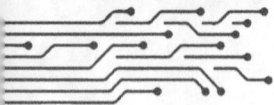

Moving on, she explained that the assessment phase encompasses three key elements, including the:

- **Core:** Understanding your organization's fundamental identity and capabilities

- **Internal Analysis:** Comprehensive review of strengths, weaknesses, metrics, benchmarks, business intelligence, and continuous quality improvement processes

- **External Analysis:** Systematic evaluation of threats, opportunities, market dynamics, venture possibilities, and innovation potential

## Phase 2: Decide

The next slide Susan presented was equally detailed and enlightening.

"Alright, so Phase 2 of the Vairos framework is Decide. This phase is what distinguishes between the insight provided by the analysis and the strategic decisions that are made based on that analysis."

"That sounds quite interesting," said Megan, "but how does all of it play out?"

"Well basically" said Susan, "In this phase, all the complex assessment conducted in the one before essentially narrows down to a few critical and focused decisions. In phase 2, one must employ a mind with both analytical powers and also one that possess some good old-fashioned leadership abilities."

Concluding for this slide, she mentioned that decision-making Vairos system involves:

- Synthesizing assessment findings into strategic options

- Evaluating alternatives against organizational capabilities and market realities

- Making clear choices about strategic direction

- Establishing decision criteria and governance processes

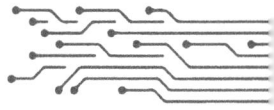

## Phase 3: Execute

Susan moved on to the third phase of the framework. "Execute," she said, " the third phase basically translates strategic decisions into measurable actions. It's what breaks broader strategic direction into specific goals, objectives, strategies, tactics, and policies. It is exactly in this phase, where you create measurable systems you can employ later for tracking progress and ensuring accountability."

This slide entailed what at all execution encompasses:

- **Planning Structure:** Goals, objectives, strategies, tactics, and policies

- **Measurement Systems:** Dashboards, evaluation processes, and feedback mechanisms

- **Governance:** Accountability structures and performance management

## The Data-Driven Difference

Perhaps what sets Vairos apart from traditional strategic planning approaches is its systematic integration of data throughout all three phases. Rather than relying solely on intuition or experience, Vairos demands evidence-based analysis at each stage.

"Susan, I find your research rather remarkable. However, I can't help but ask: what differentiates Vairos from traditional strategic planning?" Jim asked.

"That's a question I get asked a lot", Susan answered. "You see, Vairos' integration of data across all three phases is its most distinguishing feature. It mandates this critical integration, which is not evident in many other traditional methods," replied Susan. "Vairos doesn't rely on intuition." It demands evidence-based analysis at each stage. This is why success is more likely for Vairos users."

As we mulled over each new detail, Susan shifted on the next slide. What we saw next was a detailed outline of the advantages and disadvantages of this data-driven approach. They included:

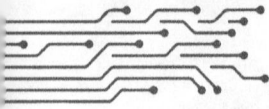

- **Objectivity:** Reduces bias in both assessment and decision-making

- **Accountability:** Creates measurable standards for success

- **Agility:** Enables rapid course correction based on performance data

- **Learning:** Builds organizational capability for continuous improvement

## Implementation Principles

The Vairos system operates on several core principles that guide implementation:

"So, I think I'm beginning to get a firm grip on this," said Megan, "Does this Vairos system come with some guidelines for implementation as well?"

"Another stellar question Megan," answered Susan, "And yes, it does come with guidelines for implementation that outline protocols for everything, from a systematic approach to the actual methodology itself."

Next, we read another detailed outline of the key principles for implementation. These principles included:

- **Systematic Approach:** Each phase builds logically on the previous, ensuring comprehensive coverage without overwhelming complexity.

- **Scalable Framework:** The system works for organizations of all sizes, from startups to established ones and adapts to organizational context while maintaining methodological rigor.

- **Iterative Process:** While the phases appear linear, Vairos operates as a continuous cycle, with insights from execution combining forces with future assessments.

- **Collaborative Methodology:** The system encourages cross-functional participation while maintaining clear accountability and decision-making authority.

"Alright, I think it's time for a little break. I believe we have sufficient information to mull over for now." said Jim. "Let's set up a meeting for next week, and pick up from where we left off. Sounds good?"

"Sure thing," replied Susan, "in our next session, I'll cover each component of the Vairos section in detail. I'm hopeful to have even more tools ready by then to aid with its implementation. Believe me when I say that Vairos will help you bridge the gap between strategy, planning, and execution at MegaHealth."

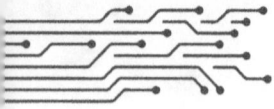

# ASSESS

Not to long after this meeting, we unanimously agreed that it was time to implement Vario's framework at MegaHealth.

This decision came on the heels of the next agenda; covering each component of the system, one by one as Susan had promised last time. It was not a surprise to see each of us returning to the boardroom with anticipated vigor.

"Alright guys, now, one of the key things you need to know is that before any organization can develop an effective strategy, it must work towards understanding its fundamental identity first. This first step is exactly what the Assess component focuses on," said Susan, as she took a breath and continued, "First things first, the fundamental is basically the essential elements that define who you are. It's what you stand for, what you believe in, how you operate, and so on."

"Yeah, but doesn't every business have that already?" asked Megan. "Um, you're right to some extent. Every business or organization has a set of fundamentals, but they're just on paper for the most part. You see, most businesses operate in a reactive mode. Along the way, organizations get so carried away with the flow and adapting with elements that they lose sight of their fundamentals. Before they realize it, they become an organization that responds to immediate pressure as priority and don't really have a clear futuristic direction."

"I know what you mean. I've seen that happen quite a lot," said Jim, "And based on what I've picked up thus far, I guess that the system likely changes organizations and pushes them towards being more proactive and efficient..."

"Yes, that's exactly what it does, but that's just the start," replied Susan.

## The Foundation Metaphor

Susan, in continuation of explaining the foundation metaphor, said, "Imagine you're working with a personal trainer. The first step you take isn't about selecting an exercise routine or setting fitness goals. It's about boosting motivation and building the core strength, right?"

"Yes, that's what trainers have always told me," Jim replied in agreement. "Hence, from what I understand you're implying that without core strength, any advanced moves you want to take on can quickly become ineffective, and even worse, dangerous. That same principle applies to organizational strategy."

## Assess Components

Next, Susan ushered us down the five critical elements of the Assess component. These elements were:

- **Mission:** The inspirational purpose that drives daily operations and speaks to the heart of why your organization exists.

- **Vision:** The aspirational future state that guides strategic direction and appeals to the imagination of what's possible.

- **Identity:** Understanding the key attributes and characteristics that define your organization's unique nature.

- **Strategic Review:** Systematic examination of past strategic efforts to understand what works and what doesn't.

- **Risk Assessment:** Honest evaluation of potential threats and vulnerabilities that could derail strategic initiatives.

## Business Model Clarity

After the core components, we then moved our attention to the business model clarity.

"Before we head into what the core definition is, you must clearly understand two fundamental concepts," said Susan. First is the business model, which in essence is a comprehensive outline of how the business operates. For the business model, you ought to have a detailed understanding of the production process, revenue drivers, and essential operating elements. But that's not all. This also investigates whether the organization is primarily a human capital-driven entity or one that favors tech. Understanding your business model is essential simply because it influences every strategic decision you make."

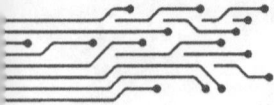

"So far so good, Susan. It makes sense. If you don't know how you're doing what you do, you can't make decisions to improve it," said Jim. "Couldn't have found better words myself. That's exactly what it is. Now, the next most important thing is the business plan. As you already know, this concerns how your organization turns ideas into revenue. It considers everything from the market opportunity analysis to venture funding, capital projects, competitive positioning, and growth strategies. " replied Susan, before moving on to the next slide.

## The Six Operating Models

We resumed our discussion after a brief pause. Susan said, "Every organization operates on one of the six primary models, and each one requires a strategic approach."

It wasn't long before Susan apprised us of the six models. They included:

1. **Partnership Model:** Highly reliant on people and expertise, common in professional services. Success in this model, depends on advancing through partnership ranks.
2. **Matrix Model:** Requires juggling multiple sides of the business simultaneously, with incentives aligned across different organizational dimensions.
3. **Diversification Model:** Spreads risk across multiple industries, products, or services, requiring sophisticated portfolio management.
4. **Assembly Line Model:** Emphasizes communication and collaboration among departments, often requiring technology tools for coordination.
5. **Parts = Sum of Whole Model:** Treats each business unit as an independent profit center, requiring strong individual accountability.
6. **Unity Model:** Focuses on single service or product excellence, emphasizing efficiency and brand development.

If you wish to develop foolproof and relevant strategic initiatives, it is imperative to completely understand your operating model and organizational structures.

## Stakeholder Analysis

"I cannot emphasize the importance of understanding a business model enough. It is one of the core elements for developing strategic initiatives- the other is stakeholders." said Susan. "Every organization, irrespective of its size, has a set of stakeholders whose interests are always priority concerns for the heads."

"I agree. It's about aligning everyone's interests in a way that they accomplish a common goal," said Megan.

"Many people often overlook this core element. You see, stakeholders are basically your customers, employees, shareholders, partners, and community members. Often organizations merely stop at identifying stakeholders; the more crucial part is to understand their relative importance and influence on the organizational success. These rankings guide strategic decision-making and resource allocation," replied Susan.

## Core Assessment Exercise

For the core assessment, you have to begin by identifying three core attributes that form the crux of your organization.

You may complete this exercise individually or in groups. However, the main idea is to capture immediate, instinctive responses rather than researched answers.

If you are conducting it as a group exercise, perhaps the following will serve as a helpful guide:

1. Encourage the participants to list down attributes of the organization on sticky notes on their own without influence from the leaders present
2. Post all responses without discussion
3. Take a break to allow an objective review
4. Organize ideas into core groupings
5. Rank/Order the Attributes of the Organization
6. Identify common themes and unique perspectives

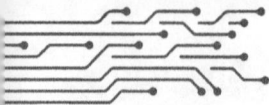

This exercise reveals how people inside the organization perceive its fundamental nature and lays the framework for the more crucial and comprehensive development work.

The Assess component establishes the bedrock upon which all other strategic elements rest. In the absence of this, even the most sophisticated analytical tools and execution frameworks will fail to deliver sustainable results.

## Mission

The next morning, MegaHealth's boardroom came alive once again, with Susan, Megan and Jim's presence.

Susan seemed to be on the most energetic side today. She opened her laptop and pulled up the first slide. "Alright," she said, "before we make our way to the more mind boggling components, let's start with something simple but powerful- the mission."

Jim leaned back in his chair. "You mean the same kind of mission statement every company has printed on the wall that nobody takes a second look at?"

Susan smiled. "Jim, this is exactly where the problem lies. The mission statement comprised of mere words, which are mostly crafted impeccably. But alas- they fail to rouse any emotions. The real mission is supposed to be the heart of the organization. It's what gives people a reason to care about what they do every day. Vision gets people dreaming, but mission inspires them and keeps them moving everyday."

Megan nodded. "So, it's less about slogans and more about energy?"

"Yes, precisely that." Susan said. "A strong mission should hit people on an emotional level. It should be a mantra that everyone in the company can repeat unthinkingly, because they believe in it. It's what keeps you grounded when things get messy or stressful."

She paused for a moment. "I'll give you an example. My church has a short mission statement, and I keep it in my wallet. Whenever there's a disagreement, I pull it out and check whether what we're doing still aligns with that purpose. It's simple, but it works." Jim raised an eyebrow. "So, you're saying the mission is like a guiding star for decisions?"

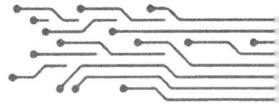

"Yeah, exactly," Susan said. "It's not just inspiration, it's direction. A good mission should remind your team why they're here. It should inspire their choices that fit into the bigger picture." Megan looked at the slide again and smiled. "Alright, that makes sense. The heart before the plan."

"Perfect way to put it," said Susan. "Without the heart, strategy doesn't last long."

## Mission

When the next slide came up, Susan looked at both Jim and Megan. "Alright, now that we know what a mission really is, let's talk about what makes one actually work."

Jim grinned. "You mean what separates a good mission from a bunch of nice empty words?"

"Exactly," said Susan. "Most companies write something that sounds impressive, but it doesn't do much for its employees. A real mission has four traits that make it stick."

She held up four fingers as she spoke.

"First," she said, "it's inspirational, not aspirational. The mission is about right now, what you're doing today, not what you hope to be in ten years. Vision is the dream. Mission is the work."

Megan nodded. "So, the mission keeps people focused on what matters in the present."

"Right," Susan said. "Second, its heart centered. It should connect with people emotionally, whether they're employees, patients, and partners or just about anyone. The mission should make their work feel significant."

Jim leaned forward. "I believe that's the part most companies tend to overlook. Their mission statement often sounds smart but lacks a touch of reality."

Susan smiled. "Exactly. Third point, a good mission is action oriented. It reflects what you do as an employee and why your work matters, not just who you want to be someday. It's about behavior, not dreams."

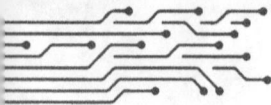

Megan looked thoughtful. "So, it's more of a guide for how we act, not just what we say."

"Yes, it is," said Susan. "And the last point- cultural alignment. The mission must match who you are as a society member. If the words don't reflect how your team behaves or what their core values are, then it's just noise."

Jim nodded. "So what you mean is that people should demonstrate the mission in the way they work. If their methodology doesn't demonstrate the mission, then it isn't realistic in the first place."

Susan smiled. "That's it. A mission only matters if people can feel it, live it, and make decisions through it."

## Examples of Effective Missions

"I'd now like to show you samples of a few missions that actually work," said Susan.

Jim smirked. "Let me guess, not the ones written by a committee of twelve?"

Megan laughed. "Definitely not."

Susan grinned. "Exactly. The first one's from Walmart. Sam Walton once said, 'The secret of successful retailing is to give your customers what they want. And really, if you think about it from your point of view as a customer, you want everything: a wide assortment of good-quality merchandise; the lowest possible prices; guaranteed satisfaction with what you buy; friendly, knowledgeable service; convenient hours; free parking; a pleasant shopping experience.'"

Jim whistled. "That's a mouthful, but it's real. You can tell he believed that."

Right," Susan said. "It's not just a slogan, it's a playbook. You can build a store, train staff, or set pricing using that sentence. It tells you exactly what matters: quality, price, service, convenience, and experience."

Megan nodded. "It's like every line is a decision filter. If it doesn't serve the customer, it doesn't fit."

"You got it," said Susan. "Now, look at this one from United Way: 'United Way improves lives by mobilizing the caring power of communities around the world to advance the common good.'"

Jim leaned back. "That's short but strong."

Susan smiled. "Yeah, it nails both the how and the why. The 'how' is mobilizing the caring power of communities. The 'why' is advancing the common good. Simple, emotional, and actionable."

Megan added, "And anyone who works there can instantly tell if what they're doing connects to that mission."

Susan nodded. "Exactly the point. The best missions aren't fancy, they're lived."

## Common Mission Statement Failures

Susan clicked to the next slide and sighed. "These were some examples that hit the mark. I now want you to look at some mission statements I consider failed attempts."

Jim squinted at the screen. "Albertsons. Okay, what's theirs?"

Susan read it out loud. "'To create a shopping experience that pleases our customers; a workplace that creates opportunities and a great working environment for our associates; and a business that achieves financial success.'"

Jim grimaced. "That sounds like something written by a lawyer who just discovered bullet points."

Megan chuckled. "It's trying way too hard to make everyone happy."

"Exactly," Susan said. "There's no focus. Customers, employees, profits, all valid things, but crammed together without any heartbeat behind them. You can't feel what they care about."

Jim nodded. "Yeah, there's no emotion. I don't know what drives the company when I read that."

"Right," said Susan. "A mission should pull you in, give you something to believe in. This one's polite, tidy, and completely forgettable."

Megan smiled. "So basically, it's a checklist, not a mission."

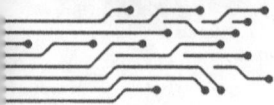

"Unfortunately, yes." Susan said, clicking to the next slide. "And that's where most companies get it wrong."

## Mission Development Rules

"The good news is that one does not need to get over ambitious while crafting a mission statement. There are only a few, simple rules to nail a mission statement", said Susan.

Jim smirked. "Let me guess, don't use big words just to sound smart?"

"Exactly," Susan said, smiling. "The very first rule is to avoid ambiguous language. Every word should convey a clear and singular meaning. If people can read it three different ways, it's not a mission, it's a riddle."

Megan nodded. "True that, clarity over cleverness always."

"Right," Susan said. "Next is to keep the structure simple. No lengthy sentences, no buzzwords. A good mission should be easy enough for people to memorize and repeat it without checking a slide."

Jim pointed at the screen. "I see, so it's more like a phrase you can say in an elevator, not a paragraph from a report."

"Yes," Susan said. "And the third rule is to use ordinary, everyday words that everyone understands. Unless you're a biotech firm or a space agency, you don't need jargon. If your front-line staff can't relate to the words, the mission won't live beyond the boardroom."

Megan smiled. "That ties right into your next point, doesn't it? Keep it simple."

"Yeah," Susan said. "If a mission's too complicated, no one remembers it, let alone believes it. The best ones would be something that people would almost tattoo on their hearts."

Jim laughed. "I'm not getting a tattoo for MegaHealth, but I get your point."

Susan grinned. "Good and the last two rules are to create an emotional connection, and make sure it aligns with how people make decisions. A mission isn't meant to decorate the website; it's supposed to steer behavior from the CEO all the way to the front desk."

Megan nodded. "So basically, if people can't feel it or use it, it's not a real mission."

Susan smiled. "Exactly. It's just fancy yet empty words."

## Mission as Decision Filter

Susan paused the presentation and looked at Jim and Megan intently, "Here's where a mission really earns its keep as a decision filter."

Jim leaned forward. "Meaning?"

"Meaning," Susan said, "when you're faced with a choice, a new product, a partnership, even a hiring decision, you should be able to ask one simple question: does this move us closer to our mission?"

Megan nodded. "And if the answer is no?"

"Then you stop," Susan said. "Or at least rethink it. If the mission doesn't support it, it's probably the wrong move."

Jim rubbed his chin. "So, it's not just something we print and frame, it's a test we run every time we take a decision for the organization."

"You got it," Susan said. "That's when the mission statement stops being wall art and starts being a management tool. Teams can use it to check if a project's worth doing, how to spend resources, or even how to settle disagreements."

Megan smiled. "Mission statement as a referee? Now that's a unique idea."

Susan nodded. "Perfect way to put it. The best missions don't just inspire, they guide."

### Mission Evolution

Susan leaned back. "Now, time for a bombshell. I know from all that we have discussed so far, you would think that missions last lifelong. In all honesty, they ought to, but just sometimes, there can be an exception. Missions may change over the course of time."

Jim raised an eyebrow. "You mean when the business shifts?"

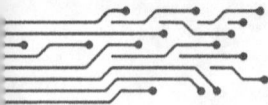

"Exactly," Susan said. "As the company grows or the market moves, your mission might need some tweaking to stay relevant. But it should never be a knee-jerk reaction to short-term problems."

Megan nodded. "Sure, it shouldn't. But how can we figure if it is time to redo our mission statement?"

Susan counted on her fingers. "Ask yourself four things: has our core purpose changed? Do new circumstances call for a different focus? Does our current mission still inspire people and guide decisions? And would changing it strengthen or weaken who we are?"

Jim smiled. "So, evolution, not reinvention."

"Yes," Susan said. "You simply adjust the wording when revamping the mission statement, but its soul, its essence, remains the same."

## Implementation Process

Susan brought up the final slide. "Alright, so here's how you actually build a long-lasting mission."

Jim leaned forward. "Let me guess, not by sending out a company-wide survey and hoping magic happens?"

Susan laughed. "You do gather input, but you don't crowdsource the final call. Start by talking to key people, leaders, long-timers, even a few customers, and get their take on what the organization really stands for."

Megan nodded. "Then you turn that into options."

"Right," said Susan. "Draft a few different versions. Don't aim for perfection yet, just capture the essence. Then test them. Ask yourself, which one hits home? Which one feels true and useful?"

Jim asked, "And once you've got a winner?"

"You refine it," Susan said. "Tighten the language and make it memorable. Once done, then communicate it everywhere, in meetings, onboarding, and performance reviews. It must live inside the organization's culture, not just on the website."

Megan smiled. "That's great. In this way, everyone gets a voice, but leadership makes the call."

"Exactly," Susan said. "It's not democracy, but it's direction through teamwork. The mission is your North Star. It keeps everyone aligned when the real work begins."

## Vision

Susan looked up from her notes. "If mission speaks to the heart, vision speaks to the mind."

Jim nodded. "So, it's about where we're going as an organization, not what we're doing right now."

"Exactly," Susan said. "Vision is the picture of the future you're working toward, the thing that fuels ambition and imagination."

Megan added, "It's the finish line everyone's running toward."

Susan smiled. "Perfect way to put it."

### The Brain Connection

Susan continued, "A good vision has to stir something; it speaks to the dreamer and the strategist at the same time."

Jim leaned back. "The vision helps people power through the hard times and stressful stuff."

"Exactly," Susan said. "Vision gives direction for the long game and the energy to keep pushing through the tough changes along the way."

Megan nodded. "It's what makes people believe the future's worth building."

## Characteristics of Effective Visions

Susan pointed to the next slide. "Here's what separates a strong vision from a pretty sentence."

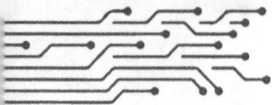

"First," she said, "it has to be aspirational but specific, big enough to stretch the imagination and clear enough to guide choices."

Jim nodded. "Not fanciful but not fixated on the past either."

"Right," Susan said. "It's future-focused, a picture of what tomorrow's organization looks like."

Megan added, "And it should make people feel eager."

"Yes, that's right," Susan said. "Hope, excitement, not just logic. And finally, it must drive strategy. If your vision doesn't shape where the money, people, and time go, it's just wallpaper."

## Examples of Effective Visions

After this, Susan moved on and shared a few examples of compelling visions. These included:

- **Ben & Jerry's:** "Making the best ice cream in the nicest possible way"

    o   This vision succeeds because it combines excellence (best ice cream) with values (nicest way) in a memorable, specific language.

- **Google:** "To provide access to the world's information in one click"

    o   This vision works because it clearly defines both the scope (world's information) and the method (one click access).

- **Coca-Cola:** "Our vision is to craft the brands and choice of drinks that people love, to refresh them in body & spirit. And done in ways that create a more sustainable business and better shared future that makes a difference in people's lives, communities, and our planet."

    o   This vision integrates product excellence with broader social impact.

- **Instagram:** "Capture and share the world's moments"

    o   Simple, clear, and emotionally resonant while defining business scope.

## Examples of Less Effective Visions

Then, she moved on to sharing some examples of less emphatic and less effective visions. These included:

- **Law Firm:** "To create a different kind of law firm, putting culture first, with great people doing the best legal work for our clients."

    o   This vision fails because it focuses on being "different" without defining what different means or why it matters.

- Dell: "Delivering a better tomorrow"

    o   Too vague to provide strategic guidance or emotional connection.

## Vision Evolution: The KFC Example

Susan said, "Even big brands miss the mark sometimes. Take KFC for instance."

Jim raised a brow. "The chicken guys?"

"Yeah," she said. "Back in 2013, their vision was all about selling fast food to health-minded, price-conscious customers. The vision's focus was purely operations oriented, but completely lacked inspiration."

Megan smiled. "So, they fixed it?"

"They did," Susan said. "The new version is 'Be Your Best Self. Make a Difference. Have Fun.' Still simple, but now it connects with people. It employs less mechanics, but more meaning."

## Mission and Vision - Development Rules Relationship

Next, Susan went on to define certain rules that one must follow for vision development:

- **Make it Short and Memorable:** Vision should be easy enough to memorize and repeat for all team members.

- **Be Organizationally Specific:** Vision should apply uniquely to your organization, not to any company in your industry.

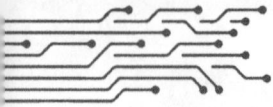

- **Be Purpose-Driven:** Vision should explain why the future state matters, not just what it includes.

- **Align with Culture and Values:** Vision must reflect the organization's authentic character, not an aspirational culture that doesn't exist.

- **Provide Strategic Guidance:** Vision should inform major decisions about resource allocation and strategic direction.

Susan continued, "Just like a mission, the vision too should guide decisions."

Jim nodded. "So, every big move needs to push us closer to that future picture."

"Exactly," Susan said. "If an idea doesn't align with the vision, it probably doesn't deserve time or money."

Megan added, "That's how you stay focused even when everything irrelevant looks like an opportunity."

Susan smiled. "Right. Mission drives what we do today. Vision defines where we're heading. Together, they keep the organization inspired, grounded, and moving in the right direction."

## Shared Vision Mistakes and Implementation Guidelines

Susan then moved our attention to some of the most common vision mistakes and implementation guidelines:

- **Too Broad:** Vision that could apply to any organization fails to provide specific guidance.

- **Too Operational:** Vision that describes current activities rather than future possibilities limits strategic imagination.

- **Too Complex:** Vision requiring paragraph-length explanation won't be remembered or used.

- **Disconnected from Reality:** Vision that ignores organizational capabilities or market realities creates cynicism rather than inspiration.

After that, she said, "Vision development follows a similar process to mission development but requires even greater attention to future possibilities."

She then shared some key items, which included:

1. **Environmental Analysis:** Understanding market trends and future possibilities
2. **Capability Assessment:** Honest evaluation of organizational potential
3. **Stakeholder Input:** Gathering perspectives on the desired future state
4. **Strategic Integration:** Ensuring vision aligns with strategic planning
5. **Communication and Embedding:** Making vision a practical tool for decision-making

Susan's concluding statement was perhaps the most enlightening and powerful takeaway from this session.

She said, "The best organizations don't treat mission and vision as posters, they use them instead."

Jim nodded. "So, they come up in real and daily conversations, not just retreats."

"Exactly," Susan said. "Vision is the North Star; it keeps long-term transformation on track. Mission is the fuel; it powers the work every single day."

Megan smiled. "Head and heart working together."

Susan nodded. "That's how great companies stay focused and alive."

## Identify

It was now time to reflect on the next section.

"Now we shift to the 'Identify phase'," Susan said as introduction. "This one is all about people, the ones who'll make or break the strategy."

Jim leaned in. "You mean identifying who actually matters and who does not as per the mission and vision?"

"Yes, exactly that," Susan said. "You figure out your key stakeholders, the real influencers, and the informal networks that are essential for

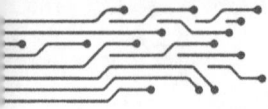

achieving objectives. Strategy doesn't fail because of ideas; it fails because you missed the people part."

Megan nodded. "So, before execution, you identify the human landscape."

"Right," Susan said. "That's where success starts."

## The Human Factor in Strategy

Susan said, "You may have the best plan in the world, but if you underscore or overlook the people essential for its success, it'll fall apart."

Jim nodded. "So even perfect strategy fails without buy-in."

"Exactly," Susan said. "Identification work makes sure you understand the politics, the hidden influential networks, and what motivates people. Strategy only works when humans do."

Megan added, "So it's not just org charts, it's understanding the undercurrents."

Susan smiled. "Right. That's the real map of execution."

### Stakeholder Mapping

Susan continued and outlined that one should always begin with systematic identification of key organizational stakeholders that include:

- **Internal Stakeholders:** Employees, managers, executives, board members, union representatives

- **External Stakeholders:** Customers, suppliers, partners, community leaders, regulatory bodies, investors

- **Influential Networks:** Industry associations, professional groups, informal advisory relationships

For each stakeholder group, one has to document:

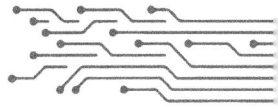

- Level of influence on strategic outcomes

- Current attitude toward changes or initiatives

- Potential contribution to strategy execution

- Specific concerns or interests that must be addressed

## The "Doers" Analysis

Susan then moved on to a concept, which she referred to as the "Doer" analysis and said, "Every company has its unofficial power players, the ones who actually make things happen."

Jim nodded. "You mean the people everyone calls when something needs fixing."

"Yes," Susan said. "They may not have titles, but they've got trust, relationships, and know-how."

Megan leaned forward. "So how do you find them?"

Susan smiled. "Ask simple questions: Who steps up in a crisis? Who organized the last big response? Who do people go to when they need something done fast? And who's missing when things fall apart?"

Jim thought for a moment. "Those answers tell you who really runs the place."

"Right," Susan said. "That's your informal power map, and it's as valuable as gold for planning execution."

## Change Champions and Resisters

"Whenever you roll out a new strategy, people split into two camps," said Susan.

Jim smirked and said "Yes, the ones who lean in and the ones who dig in."

"Some see change as a chance to grow; others see it as a threat. You must understand both if you want the plan to stick," replied Susan.

"So, managing strategy starts with managing reactions." Susan said.

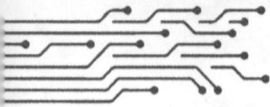

"It's as much psychology as it is planning."

Susan proceeded to outline some of the key characteristics for both champions and resisters of change. These included:

**Change Champions** typically:

- Demonstrate comfort even in ambiguity
- Show track record of adaptation
- Possess influence over key stakeholder groups
- Express dissatisfaction with current state

**Change Resisters** typically:

- Benefit from current systems and processes
- Fear loss of status, authority, or job security
- Lack confidence in organization's change capability
- Have been burned by previous failed initiatives

Susan said, "I wouldn't blame you for being prejudiced in this case. But, here's the key: neither group is wrong."

Jim raised an eyebrow. "Even the ones fighting the change?"

"Yeah," Susan said. "Their resistance usually makes sense once you understand their situation. The trick is to address their concerns without destroying the energy of those ready to move." Megan nodded. "So, you build trust with one side and momentum with the other."

"I agree", said Megan. "That's how real change takes hold."

## Keystone Habits and Cultural Anchors

Both Jim and Megan appeared quite convinced, so Susan continued, "Organizations develop habits and patterns of behavior that become automatic and self-reinforcing. Some habits support strategic goals; others undermine them. Understanding these patterns helps predict implementation challenges and opportunities."

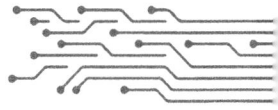

On the slide she had written:

**Keystone Habits** are behaviors that naturally trigger other positive behaviors. Examples might include:

- Regular performance review meetings that drive accountability
- Data-driven decision making that improves strategic choices
- Customer feedback processes that focus organizational attention
- Innovation time that generates new possibilities

**Cultural Anchors** are the informal rules and traditions that shape daily behavior:

- "How we really make decisions around here"
- "What gets rewarded and what gets punished"
- "Which meetings matter and which are just theater"
- "Where the real work happens"

## Assessment Process

After that she moved on to the next slide and continued with outlining an assessment process. This included:

- Interviews: One-on-one conversations with key stakeholders about organizational dynamics, past initiatives, and change readiness
- Observation: Attending meetings, reviewing communication patterns, and observing informal interactions
- Network Analysis: Mapping relationships and influence patterns through surveys or organizational chart analysis
- Historical Review: Examining past change initiatives to understand what worked, what failed, and why

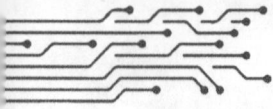

## Capacity and Capability Analysis

Then Susan continued and said, "Identify both current capacity (available resources) and capability (skills and competencies) for strategic execution." She again outlined a few key points that included:

- **Leadership Capacity:** Do key leaders have bandwidth for strategic initiatives, or are they overwhelmed with operational demands?

- **Skill Capabilities:** Does the organization possess technical skills, project management capabilities, and change management competencies required for strategy execution?

- **Resource Availability:** Are financial, human, and technological resources adequate for strategic initiatives?

- **Cultural Readiness:** Does organizational culture support the level of change and collaboration required for strategic success?

## Political Landscape

After that she went on to talk about the political landscape and stated, "Every organization has political dynamics that influence strategic outcomes."

Understanding these dynamics allows for more effective navigation:

- Formal Authority: Who has decision-making power for resources and strategic direction?

- Informal Influence: Who shapes opinions and influences decisions without formal authority?

- Coalition Potential: Which groups might naturally align around strategic initiatives?

- Conflict Areas: Where are the likely sources of resistance or disagreement?

## Documentation and Application

Then, she shared one of the most important aspects of the process and said that professionals should always "Create stakeholder maps, influence diagrams, and capability assessments that inform strategy development and implementation planning."

She mentioned that the documentation should include:

- Stakeholder analysis matrix with influence/interest ratings
- Change readiness assessment by organizational level
- Capability gap analysis with development recommendations
- Political landscape mapping with navigation strategies

## Review Strategies

Susan moved to the next point. "Before you chart onto a new direction, it is good to first look back and analyze the ground you have covered."

Jim nodded. "Figure out what actually worked and what didn't."

"Exactly," Susan said. "The Review Strategies phase is about studying your past moves, what succeeded, what fell apart, and why. That's the intelligence you need before making the next big call."

Megan added, "So it's not just reflection; it's research for smarter decisions."

Susan smiled. "Right. History is data, if you treat it that way."

## Learning from Strategic History

One of the key insights Susan shared was that most organizations repeat strategic mistakes because they fail to systematically analyze their strategic track record.

The 'Review Strategies' component creates organizational memory about strategic effectiveness. This in turn helps leaders make more informed decisions about future initiatives.

With the help of this review, one can examine multiple dimensions of past strategic performance. These include:

- **Outcomes**: Did strategies achieve intended results?
- **Process**: How well did planning and implementation work?

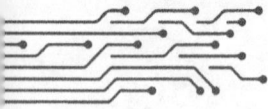

- **Context:** What environmental factors influenced results?
- **Learning:** What insights can inform future strategy?

## Strategic Initiative Inventory

Susan mentioned that Jim and Megan should "begin by cataloging strategic initiatives from the past three to five years."

She pointed out that the inventory could include things like:

- **Major Strategic Plans:** Comprehensive organizational strategies, five-year plans, major transformational initiatives
- **Significant Projects:** Large investments in technology, facilities, market expansion, new service lines
- **Operational Improvements:** Process redesign, efficiency initiatives, quality improvement programs
- **Cultural Changes:** Leadership development, organizational restructuring and value implementation programs

For each initiative, document:

- Original objectives and success metrics
- Resources invested (financial, human, time)
- Actual outcomes and performance against goals
- Implementation challenges and unexpected obstacles
- Key decisions that influenced results

## Success and Failure Factor Analysis

"When you review past strategies, look for patterns in what actually worked", said Susan.

Jim asked, "Like what kind of patterns?"

"Things like strong leadership support, clear communication, or quick early wins," Susan replied. "You'll start seeing themes regarding the conditions that made success possible."

Megan nodded. "So, we're not guessing what works next time, we're repeating what's proven."

According to Susan, some of the key factors worth considering for such an analysis include:

**Strategic Design:** What characteristics made strategies more likely to succeed?

- Clear, measurable objectives
- Realistic resource requirements
- Strong stakeholder alignment
- Appropriate timing and sequencing

**Implementation Excellence:** What execution factors drove success?

- Strong project management
- Effective communication
- Adequate resource allocation
- Consistent leadership support

**Environmental Conditions:** What external factors supported success?

- Market timing
- Competitive dynamics
- Regulatory environment
- Economic conditions

Susan believes that in addition to analyzing past successful patterns, one must also examine unsuccessful initiatives to understand failure patterns:

**Strategic Flaws:** What design problems caused difficulties?

- Unclear objectives or success metrics
- Unrealistic resource assumptions

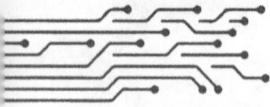

- Poor stakeholder analysis
- Inadequate risk assessment

**Execution Problems:** What implementation issues derailed progress?

- Insufficient project management
- Poor communication and change management
- Resource shortfalls or competing priorities
- Leadership turnover or loss of support

**External Disruptions:** What environmental changes created problems?

- Market shifts or competitive responses
- Regulatory changes
- Economic disruption
- Technology evolution

## Resource Allocation Patterns, Decision-Making, Cultural and Competitive Intelligence

For the next hour or so, Susan guided our attention through numerous slides. She also talked at length about some key insights about resource allocation, cultural decision-making and more.

Some of these insights included:

**Resource Allocation:**

- **Investment Distribution:** Where did the organization invest most heavily? Which areas received adequate resources versus inadequate support?
- **ROI Analysis:** Which strategic investments generated positive returns? Which failed to meet expectations?
- **Opportunity Cost Assessment:** What opportunities were missed due to resource allocation to other initiatives?
- **Capacity Constraints:** When did resource limitations constrain strategic effectiveness?

### Decision-Making Effectiveness:

- **Decision Speed:** How quickly did the organization make strategic choices? Were delays costly or beneficial?

- **Information Quality:** What information was available for strategic decisions? What information was missing or inaccurate?

- **Stakeholder Involvement:** Who participated in strategic decisions? Were the right voices included?

- **Follow-Through:** How well did the organization execute strategic decisions once made?

### Organizational Culture:

- **Capability Development:** Which strategies built lasting organizational capabilities? Which failed to create sustainable improvements?

- **Cultural Impact:** How did strategic initiatives change organizational culture, values, and behaviors?

- **Knowledge Transfer:** What strategic knowledge was captured and retained? What learning was lost?

- **Change Readiness:** How did past experiences influence organizational appetite and potential for future strategic change?

### Competitive Intelligence:

- **Market Position:** How did strategies affect competitive position and market share?

- **Differentiation:** Which strategies created sustainable competitive advantages? Which failed to differentiate effectively?

- **Competitive Response:** How did competitors react to strategic initiatives? Were responses anticipated?

- **Industry Evolution:** How well did strategies position the organization for industry changes?

### Strategic Review Methodology

"A proper strategy review isn't guesswork, it's a structured process", began Susan.

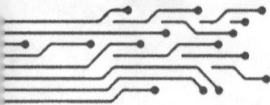

Jim leaned forward. "Walk me through it."

"First," Susan said, "you collect the data, reports, metrics, and feedback from the people who lived it. Then you bring the key leaders together for real conversations about what worked and what didn't."

Megan nodded. "And from there, you look for patterns."

"Right," Susan said. "You connect the dots, pull out the lessons, and turn them into clear insights and recommendations. Then you document it all; from performance summaries, success factors, failure points, resource use, decision quality to everything."

Jim smiled. "So, it's not merely about history; it's about using the latter to plan smarter."

"Yes, it is," Susan said. "That's how you turn experience into strategy instead of repeating the same mistakes."

**Assess Risk**

"Now comes the reality check, risk assessment", said Susan.

Jim nodded. "Is this where we stop dreaming and start testing?"

"Pretty much," Susan said. "Strategy is usually about opportunities, but you can't ignore what could go wrong. You've got to look at the threats, weak spots, and failure points that could derail everything."

Megan added, "From how I perceive it, you could describe it as optimism with a seatbelt." Susan smiled. "Exactly, smart strategy plans for both success and survival."

Soon after, Susan announced that it was time to end for the day. Before we wrapped up, she went over the last slide that covered strategic risk, assessment methodologies, and more.

Some of the key insights from her presentation, on this concept, included:

## Strategic Risk Categories

- **Market Risks:** Changes in customer demand, competitive dynamics, or industry structure that could undermine strategic assumptions

- **Operational Risks:** Internal capacity constraints, capability gaps, or execution failures that could prevent strategy implementation

- **Financial Risks:** Resource shortfalls, cost overruns, or revenue disappointments that could force strategic modifications

- **Regulatory Risks:** Policy changes, compliance requirements, or legal challenges that could block strategic initiatives

- **Technology Risks:** System failures, cyber threats, or technological obsolescence that could disrupt operations

- **Reputation Risks:** Brand damage, stakeholder conflicts, or problems in public relations that could undermine strategic credibility

## Risk Assessment Methodology

- **Risk Identification:** Comprehensive catalog of potential threats to strategic objectives, developed through brainstorming, scenario planning, and historical analysis

- **Probability Assessment:** Realistic evaluation of how likely each risk is to occur, based on historical data, expert judgment, and environmental analysis

- **Impact Analysis:** Assessment of potential consequences if risks materialize, including financial costs, strategic delays, and organizational disruption

- **Risk Prioritization:** Focus attention on the highest-priority risks based on a combination of probability and impact

## Strategic Vulnerability Analysis

- Assumption Dependence: Which strategic assumptions, if wrong, would invalidate the entire approach?

- Single Points of Failure: What individual elements, if they fail, would derail strategic progress?

- Resource Concentration: Where is the organization investing heavily without diversification or backup plans?

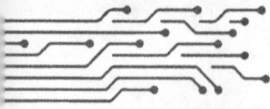

- Timeline Dependencies: Which strategic elements must occur in a specific sequence, creating vulnerability to delays?
- Stakeholder Dependencies: Which key stakeholders, if they withdraw support, would compromise strategic success?

## Competitive Risk Assessment

- **Competitive Retaliation:** How might existing competitors react to strategic moves? What defensive or offensive responses should be expected?
- **New Entrant Threats:** Could strategic initiatives attract new competitors or make market entry more attractive to outsiders?
- **Supplier/Partner Risks:** How might key business relationships change in response to strategic initiatives?
- **Customer Reaction:** Could strategic changes create customer dissatisfaction or defection?

## Scenario Development

- **Best Case Scenario:** Everything goes according to plan—what would success look like in this case?
- **Most Likely Scenario:** Realistic expectations based on organizational history and market conditions
- **Worst Case Scenario:** Multiple problems coincide—how would the organization respond?
- **Black Swan Events:** Low-probability, high-impact events that could fundamentally change the strategic context
  - For each scenario, one ought to evaluate the following points:
    - How would strategic initiatives perform
    - What modifications might be required
    - Whether the organization could adapt effectively
    - What early warning signs might provide advance notice

## Risk Mitigation Planning

- **Risk Avoidance:** Modifying strategic approaches to eliminate specific risks entirely

- **Risk Mitigation:** Reducing the probability or impact of risks through preventive measures

- **Risk Transfer:** Using insurance, partnerships, or contracts to shift risk onto other parties

- **Risk Acceptance:** Acknowledging risks that cannot be economically managed and planning for potential consequences

- **Contingency Planning:** Developing specific response plans for high-impact risks

## Early Warning Systems

- **Key Risk Indicators:** Metrics that provide early warning of potential problems

- **Environmental Scanning:** Systematic monitoring of external changes that could affect strategic initiatives

- **Stakeholder Feedback:** Regular input from key stakeholders about emerging concerns or problems

- **Competitive Intelligence:** Ongoing monitoring of competitive actions and market changes

## Organizational Risk Capacity

- **Risk Tolerance:** How much uncertainty can the organization handle while maintaining operational effectiveness?

- **Crisis Management Capability:** Does the organization have systems and leadership for managing major problems?

- **Financial Resilience:** Are financial resources adequate to weather strategic setbacks or unexpected costs?

- **Learning Capability:** Can the organization adapt quickly when strategic assumptions prove incorrect?

## Risk Communication

- Leadership Alignment: Ensure the leadership team shares a realistic understanding of strategic risks

- Board Reporting: Provide governance oversight with appropriate risk information

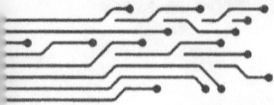

- **Stakeholder Communication:** Help key stakeholders understand the various risk management approaches

- **Team Preparation:** Prepare implementation teams to recognize and respond to risk indicators

**Integration with Strategic Planning**

- **Strategy Development:** Consider risks when evaluating strategic alternatives

- **Resource Allocation:** Factor risk management costs into strategic investment decisions

- **Implementation Planning:** Build risk mitigation systems

# DECIDE

The Decide phase transforms strategic analysis into strategic action by making clear choices about organizational direction and priorities.

Effective decision-making combines analytical rigor with leadership judgment to select strategies that advance organizational objectives. Simultaneously, the strategies also strive to remain realistic about implementation challenges and resource constraints.

## Decision Science Integration

Moving on, Susan said, "The Decide phase is where strategy meets discipline; its focus is on making smarter choices."

Jim asked, "Smarter how?"

"By using decision science," Susan replied. "You base decisions on solid data, test them through scenarios, and keep track of why you chose what you did. That way, you can learn later."

Megan added, "And you watch out for bias, right?"

"Exactly," Susan said. "You stay aware of blind spots, get diverse input, but keep decision authority clear. Good strategy isn't luck, it's structured thinking."

## Behavioral Economics in Strategic Decisions

After that, Susan began discussing strategic decisions and said, "Here's the part most leaders overlook; decisions aren't just logic, they're psychology."

Jim raised an eyebrow. "You mean emotions sneak into strategy?"

"Every time," Susan said. "You've got to know your organization's risk tolerance and how much uncertainty people can handle. At the same time, you also have to keep watch for traps like loss aversion, where you fear losing more than you value winning."

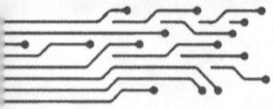

Megan nodded. "Also anchoring, like getting stuck on the first piece of info you hear."

"Yes, that's right," Susan said. "Add confirmation bias and groupthink to the mix, and you've got trouble. The best decision-making culture encourages dissent, challenges assumptions, and keeps emotion in check."

## Decision Process Design

Moving on to the decision process design, Susan leaned forward. "Good strategy isn't about a single brilliant idea; it's about choices. Real, deliberate choices. The first step is to pull together everything you've learned from the assessments, market data, internal reports, stakeholder feedback, and translate it into insights that inform action. Not just numbers, but patterns, risks, and opportunities."

She paused, tapping her pen. "Then comes the part most teams skip- generating alternatives. Don't fall in love with the first idea that sounds good. Build three or four distinct ways forward, each with its own logic and trade-offs. Strategy isn't a checklist; it's an exercise in comparison."

Jim nodded. "So, once we've got the options, we just pick the best one?"

"Not quite," said Susan. "You apply decision criteria that everyone agrees on beforehand, things like alignment with long-term goals, financial impact, and execution feasibility. It's not about intuition alone. Criteria keep the process objective, especially when the discussion gets heated."

She leaned back. "And yes, consult key stakeholders, finance, operations, even external partners, if they'll be affected. But remember, consultation isn't the same as consensus. You listen broadly but decide narrowly."

Finally, she added, "And one last thing, document everything. The rationale, the assumptions, what you expect to happen. Because when results start to shift six months later, that record becomes your anchor. It turns hindsight into learning instead of blame."

She smiled. "That's structured decision-making. Not bureaucracy, but discipline."

## Leadership Decision-Making

Susan looked around the table. "There comes a point when all the analysis in the world stops helping. That's where judgment begins. Strategy isn't just data, it's deciding what really matters."

She underlined Vision on the whiteboard. "Every decision should pull the organization closer to its long-term vision. If it doesn't, it's just noise: loud but meaningless."

Jim raised his hand. "What about instances when different teams want different things? Finance says one thing, operations another?"

"Welcome to leadership," Susan said, half-smiling. "You'll never make everyone happy. The goal isn't harmony, it's balance. You listen, weigh the trade-offs, and make a call that keeps the mission intact."

She drew a circle labeled Resources. "And keep it real if you make decisions with the resources, like money, time, and people. A bold idea that can't be executed is just theater."

Jim nodded. "So, we go for what's doable?"

"Not just doable," Susan said. "Directionally right. The decision should stretch the organization without snapping it."

She leaned back. "And always factor in your team's change capacity. You can't lead a transformation with a team already running on fumes."

Susan paused, letting it sink in. "Analysis gives you the map. Judgment tells you which road to take and how fast you can drive."

## Decision Timing and Sequencing

Susan leaned back in her chair. "Timing," she said, "is the part everyone underestimates. You can have the right strategy, the right people, the right plan, but if the timing's off, it all falls apart."

Jim crossed his arms. "You mean market timing?"

"That's one piece," Susan replied. "You've got to know when the

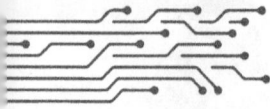

market's open to a move, when competitors are distracted, when trends are shifting, when customers are ready to listen. But it's not just about the market. It's also about you at the same time."

She pointed at the whiteboard. "Organizational readiness. Can your team deliver? Do you have the systems, the leadership alignment, and the energy to execute? If not, even a great idea will stall."

Jim nodded slowly. "And I'm guessing money matters too."

Susan smiled. "Always. Resource availability, budget cycles, capital, and even bandwidth. You must line decisions up with when the resources exist, not when the PowerPoint says they will."

She paused. "And don't forget dependencies. Some moves can't happen until others do. Miss that sequence and you create chaos."

Jim looked thoughtful. "So, it's not just about moving fast, it's about moving right."

"Yes, that's right!" Susan said. "Every big move has a window of opportunity. The trick is to see it early enough and act before it closes."

## Decision Validation and Testing

Susan turned to the group and said, "Before we all vote in complete favor of a decision, we need proof that it actually works."

Jim frowned. "Proof as in data?"

"Data, experience, and feedback," she said. "Start with a pilot program. Test the idea on a small scale before betting the farm. It's faster to fix mistakes when they're still cheap."

Megan added, "And what about the people side?"

Susan nodded. "You validate with the people who'll live with the decision: employees, customers, and partners. If they don't buy in, execution will stumble no matter how good the plan looks on paper."

She paced a little. "Then check your competitive blind spots. How might rivals react? Will this trigger price wars, copycats, or defensive plays?"

Jim leaned forward. "And the money side?"

"Run the models," Susan said simply. "Stress-test the numbers. What happens if costs rise or demand drops? That's where risk assessment comes in. You don't want to discover fragility after launch."

She smiled softly. "Validation isn't hesitation. It's basically insurance. You test, you learn, and then you scale with confidence."

## From Decision to Execution Planning

Susan picked up a marker and drew a line on the whiteboard. "This," she said, "is where strategy becomes real. You can't stop at the decision; you've got to turn it into execution."

Jim smiled. "So, the talking ends here?"

"Pretty much," she said with a grin. "Start with goals, clear and measurable ones. Not 'grow the business,' but how much, by when, and who's accountable."

Megan leaned in. "And resources?"

"Exactly," Susan replied. "Decisions don't execute themselves. You align people, budgets, and tools with what you just agreed to do. If the resources don't match the ambition, the plan dies on day one."

She wrote a timeline on the board. "Then build a realistic timeline. Set milestones so you can see if you're moving forward."

Jim nodded. "And someone owns each part."

"Right," said Susan. "Accountability isn't about blame, it's about clarity. Everyone should know what's theirs."

She capped the marker. "Finally, measure it. Set performance metrics that show whether your strategy is working. Because if you can't track it, you can't manage it, and if you can't manage it, it's just a wish list."

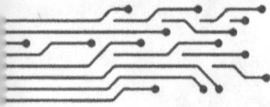

## Decision Communication

Susan looked around the table. "Here's where most strategies die: communication," she said flatly. "You can make a brilliant decision, but if people don't understand why it was made, they won't back it."

Jim raised an eyebrow. "So, we just need better messaging?"

"Not messaging," Susan corrected. "Explanation. People need to hear the reasoning, the trade-offs, the data, and the intent. Otherwise, they fill in the blanks with rumors." Megan nodded. "And we also have to be clear about what it won't do, right?"

"Yes, that's right," said Susan. "Set expectations early. Overpromising kills credibility."

She continued, "Then there's role clarity. Every department needs to know how the decision affects them. Who's responsible, who's impacted, and what's changing."

Jim leaned forward. "And I guess that's where change management comes in."

Susan smiled. "Right again. You prepare people, emotionally and practically, for what's coming. And once it's rolling, you build feedback loops so you can adjust in real time."

She paused, then added, "Communication isn't a one-time announcement. It's the oxygen that keeps strategy alive."

## Common Decision-Making Failures

Moving on to some common decision-making failures, Susan said, "You know what kills more strategies than bad ideas? Decision traps."

Jim looked curious. "Like what?"

"Well," she began, "first there's analysis paralysis, drowning in data, and never actually deciding. Teams keep researching, keep refining, but never pull the trigger."

Megan chuckled. "I've seen that movie."

"Then there's the opposite," Susan said. "Premature closure. Jumping to a decision before testing assumptions or exploring alternatives. It feels efficient, until it backfires."

Jim leaned back. "And I'm guessing optimism makes the list."

"Absolutely," Susan nodded. "Resource optimism, assuming we'll have more time, money, or people than we really do. It's how execution plans fall apart."

She pointed her pen toward the board. "Don't forget complexity underestimation. Thinking of a strategy is straightforward when it's a web of dependencies. Miss one link, and the whole chain snaps."

"And finally," she added, "Stakeholder misalignment. Making a decision that sounds great on paper but has no real buy-in. Without support, even the best strategy stalls."

Megan smiled wryly. "So basically, think hard, but not forever. Decide smart, but not blind, and make sure the people actually want to do it."

Susan grinned. "Exactly. That's how you keep strategy moving instead of spinning."

## Continuous Quality Improvement (CQI)

Susan tapped her pen against the table. "Continuous Quality Improvement isn't about big, flashy fixes," she began. "It's about creating a rhythm; small, consistent refinements that make the organization stronger every single day."

Jim nodded. "Does that mean it's not just a one-time performance push?"

"Exactly," Susan said. "It's the habit of asking: How can we make this better? Whether it's a customer process, an internal system, or team communication, improvement must be ongoing."

Megan leaned forward. "But that can get tiring. How do you keep people motivated?"

"You make it part of the culture," Susan replied. "You celebrate

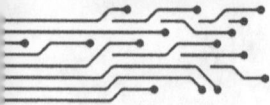

progress, even the small wins. You show people that quality isn't someone else's department, it's everyone's responsibility."

She paused. "When CQI becomes second nature, it stops feeling like a program and starts feeling like pride in your own work."

## Venture and Innovation

Susan turned the page in her notebook. "Now, innovation," she said, "is where courage meets curiosity."

Jim smiled. "You mean the part where we take risks."

"Yes," she said, "but smart risks. Innovation isn't chaos, it's structured exploration. You test ideas, run pilots, learn fast, and scale what works."

Megan added, "So, not every idea has to be a home run?"

"Yes, that's right." Susan nodded. "In fact, most innovations start small. The trick is creating a space where people feel safe to experiment, and to test something new without fearing failure."

Jim leaned back. "So, improvement keeps us stable, and innovation pushes us forward."

"Perfect summary," Susan said. "Continuous improvement protects the present. Innovation creates the future. The balance between the two is where real strategic growth happens."

# EXECUTE

Moving on, Susan began discussing the Execute phase, which translates strategic decisions into measurable action through systematic implementation planning and performance management.

This phase is the bridge between strategic intention and operational reality, ensuring that strategic choices drive actual organizational behavior and outcomes

## From Decision to Discipline

Susan stood by the whiteboard and wrote one word across the top: Execution.

"This," she said, "is where all the planning either becomes real, or dies in a binder."

Jim chuckled. "So, we're out of the theory phase?"

"Completely," Susan replied. "Execution is about turning ideas into motion. You take those big strategic concepts and break them down into pieces people can perform, measure, and manage."

She drew a quick outline. "Think of it as an execution architecture. It starts with goals, the big outcomes you want to achieve. Then come objectives, which are the measurable checkpoints along the way."

Megan nodded. "And strategies?"

"Those are the broad approaches, the how behind reaching the goals. Tactics," she continued, "are the boots-on-the-ground actions that make those strategies happen."

Jim looked at the list. "And policies?"

"Policies keep execution consistent," Susan said. "They're the rules and boundaries that keep everyone aligned. Without them, even great execution turns into chaos."

She stepped back from the board. "So, the structure's simple: goals

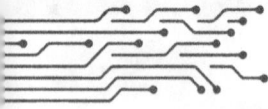

define direction, objectives show progress, strategies chart the route, tactics do the driving, and policies keep you on the road."

## Strategic Focus and Prioritization

Susan leaned forward. "Here's the truth most leaders hate to admit," she said. "You can't do everything at once. Most organizations can only execute three to five core strategies effectively. Anything beyond that, and you start trading quality for chaos."

Jim raised an eyebrow. "So, focus isn't just a preference, it's a survival rule?"

"You nailed it," Susan said. "It's what I call selective excellence. You pick fewer things and do them brilliantly instead of half-doing ten."

Megan nodded. "That also means putting real weight behind what matters most."

"Right," Susan agreed. "That's resource concentration. You give each initiative the time, budget, and talent it needs to win, rather than just exist."

Jim leaned back. "And capability alignment?"

Susan smiled. "You play to your strengths. Don't chase shiny opportunities your team isn't built for. Match initiatives to what you're already great at; that's how you gain traction fast."

"And timing," Megan added. "That has to matter too."

"It does," Susan said. "Sequential execution and rolling out strategies in waves so the organization can breathe. Focus isn't about doing less forever. It's about doing what matters now, so you're strong enough for what's next."

## The Execution Hierarchy

Susan drew a pyramid on the whiteboard. "Execution falls apart when people can't see how their daily work connects to strategy," she said. "You need a clear hierarchy that links the big picture to the smallest task."

Jim pointed at the top of the pyramid. "So, it starts here?"

"Right," Susan nodded. "At the strategy level, the broad direction, the overall play for how you'll win. That's where competitive advantage is defined."

She moved down the diagram. "Then you've got goals, measurable results that show whether you're actually moving in the right direction."

Megan added, "And objectives are the steps under that?"

"Exactly," Susan said. "Objectives are the specific milestones, the deliverables within a set timeframe. They make progress visible."

She tapped the next layer. "Then come the tactics, the actual projects and activities that make the plan move. This is where real work happens."

Jim squinted at the base. "And policies?"

"They're the guardrails," Susan said. "The principles and rules that keep everyone executing in the same direction. Without that hierarchy, daily actions drift, and strategy becomes a slogan instead of a system."

## Strategic vs. Operational Goals

Jim leaned back in his chair. "So, what's the real difference between strategic and operational goals? Aren't they both about hitting targets?"

Susan smiled. "Not quite. Strategic goals are about transformation, changing where and how the organization competes. Think of market expansion, developing new capabilities, or repositioning ourselves against competitors."

Megan nodded. "I understand those are the big, forward-looking moves. But what about the operational goals?"

"Those keep the engine running," Susan explained. "Efficiency improvements, maintaining quality, managing costs; basically, all the things that preserve current performance."

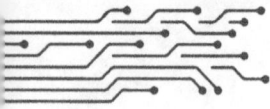

Jim frowned. "So, we need both?"

"Absolutely," Susan said. "But here's the trap: most organizations spend too much time optimizing operations and not enough on advancing strategy. The real leverage comes from emphasizing strategic goals that shift our position in the market, not just protect the status quo."

## Goal Development Methodology

Jim was scribbling notes when Megan looked up from the whiteboard. "Okay, so how do we make sure our goals work? Like, how do we know they connect back to strategy?"

Susan pointed to the screen. "That's where strategic alignment comes in. Every goal should tie directly to the organization's mission and direction; otherwise, it's just noise."

Jim nodded. "Right. And we use the SMART thing, right?"

"Exactly," Susan said. "Specific, Measurable, Achievable, Relevant, and Time-bound. It's not just a checklist, it's how we make goals real and trackable."

Megan added, "And we focus on outcomes, not just activities. It's not about how many meetings we held, it's about what changed because of them."

Susan smiled. "Perfect. And don't forget stakeholders. Our goals must create real value for them. "But" she paused, "we also must be realistic." Ambition is great, but goals need to fit within the resources we have or can secure."

Jim grinned. "So basically, align, measure, focus on results, add value, and stay grounded."

"Yes, that's right," Susan said. "That's goal design done right."

## Objective Setting and Management

Jim leaned back in his chair. "Okay, so we've got our goals, but how do we make sure they move forward? Goals are big and abstract; we need something more tangible."

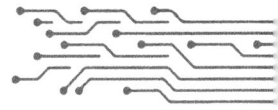

Megan nodded. "That's where objectives come in. They're like the building blocks; smaller, concrete targets that help us measure progress and keep people accountable."

Susan clicked through the slide deck. "Exactly. Each objective needs a milestone —a specific achievement that shows we're on track. And every milestone must sit on a realistic timeline, accounting for dependencies and constraints."

Jim scribbled on his notepad. "So, who's responsible for what?"

"Good question," Susan replied. "That's responsibility assignment; every objective needs a clear owner. No shared ownership, no ambiguity."

Megan added, "And let's not forget resources. People can't hit targets if they don't have the tools, budget, or time."

Susan nodded. "Right. And once everything's in motion, we keep progress tracking consistent, maintain regular check-ins, and provide reports and dashboards. It's how we stay proactive instead of reactive."

Jim smiled. "So, objectives turn strategy into motion, and accountability keeps it in motion."

## Strategy Implementation

Jim stood by the whiteboard, drawing a triangle labeled "Strategy."

"Here's the thing," he said. Goals tell us what we want. Strategy tells us how we'll get there."

Megan leaned forward. "You mean different approaches for different situations, right?"

"Exactly," Jim replied. "Take a Differentiation Strategy, which enables standing out. Creating a unique value that makes people choose us over anyone else."

Susan added, "And if we're not differentiating, we might go for Cost Leadership, being the most efficient player, delivering quality at lower cost. It's not glamorous, but it wins markets."

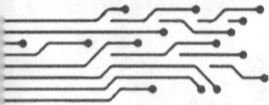

Megan added, "Then there's the Focus Strategy, narrowing in on a specific customer segment or niche. You don't compete everywhere; you dominate somewhere."

Jim nodded. "And for companies like ours that thrive on change, Innovation Strategy is key. This involves finding new products, services, or capabilities that rewrite the rules."

Susan emphasized the last point. "Don't forget Partnership Strategy. Sometimes the smartest move isn't to do it alone. You create advantage by teaming up with others who fill your gaps."

Megan smiled. "I believe strategy is really just disciplined creativity, choosing how to win, not just where to play."

Jim capped his marker. "Exactly. Strategy is the bridge between ambition and action."

## Tactical Planning and Management

Jim leaned back in his chair and said, "Alright, big strategies are great, but tactics are where we actually move the needle."

Susan nodded. "You got it. Tactics are the boots on the ground, the day-to-day actions that make strategy real."

Megan glanced at her notes. "So, this is where Project Management comes in, right? Making sure every initiative has a clear plan, milestones, and accountability."

"Right," Jim said. "And then there's Resource Coordination, we've got to align our people, budgets, and tools so we're not fighting for the same resources."

Susan added, "Timeline Management keeps everything synchronized. If one tactical stream lags, the whole strategy slows down."

Megan looked thoughtful. "And Quality Assurance makes sure we don't just finish projects; we finish them right. Execution without quality is wasted effort."

"Yes," Jim replied. "And finally, performance monitoring. We track

results, spot issues early, and adapt. Tactics without feedback loops turn into guesswork."

Susan smiled. "So, strategy gives direction, but tactics give traction."

Jim grinned. "Perfectly said. Tactics are where vision turns into visible progress."

## Policy Development and Implementation

Susan looked up from the whiteboard and said, "Policies often get a bad reputation, but they're not about control in the first place; they're really about consistency. They're what keep strategy execution from turning into chaos."

Jim nodded. "So basically, they're those guardrails you mentioned?"

"Yes," Susan said. "Behavioral Guidelines define how people should act when carrying out a strategy, not just what to do, but how to do it. It sets the tone for professional judgment."

Megan added, "Then there are Decision Frameworks, which decide what, and how. Without that, execution gets stuck in approval loops or worse, conflicting calls."

Jim scribbled something on his notepad. "And Resource Allocation policies stop us from wasting money and manpower at every shiny new idea. Keeps spending aligned with priorities."

Susan smiled. "Right. And don't forget Quality Standards; they're the essential elements that help separate disciplined execution from improvisation. Everyone knows the bar they need to meet."

every piece of execution stays in line with the bigger picture."

## Execution Monitoring and Control

Megan walked the team through the large digital dashboard projected on the wall. "This," she said, "is where execution becomes visible."

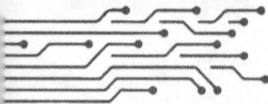

Jim leaned in. "So, every initiative gets tracked here?"

"Yes," Megan replied. "The Performance Dashboards show real-time progress, milestones achieved, resources consumed, timelines slipping or holding. It's not just data; it's the story of execution unfolding."

Susan added, "But dashboards only work if we use them. That's why we hold regular reviews; not to assign blame, but to understand what's working and what's not."

Jim nodded. "And if something's off?"

"That's where Course Correction comes in," Megan said. "We don't wait until the end of the year to react. We adapt early, when it still matters."

Susan leaned forward. "And remember, Accountability Systems must be real; people need to know their own outcomes, not just activities."

Megan concluded, "And after all that, we do Learning Integration, capturing what we've learned and feeding it back into the next cycle. Because strategy isn't static. The smartest organizations turn execution experience into future advantage."

Jim smiled. "So, the dashboard isn't just a tracker, it's a feedback engine."

"Yes," Megan said. "Execution without learning is just motion. We're after progress."

## Change Management for Execution

Susan stood by the whiteboard, marker in hand. "Alright," she said, "this is the part that trips most organizations, not the planning, but the change that comes with it."

Jim crossed his arms. "You mean the people side of strategy?"

"Yes," Susan replied. "Every strategic shift needs a Communication Strategy. People don't resist change as much as they resist confusion. You've got to tell them what's changing, why it matters, and how it affects their world."

Megan added, "You shouldn't forget Training and Development. We can't expect teams to execute a new strategy using old tools or old mindsets. Capability-building must move in sync with execution."

Susan nodded. "Right. Then there's Cultural Alignment; strategy fails fast when the culture pulls in the opposite direction. You can't say innovation matters if your systems punish risk."

Jim smirked. "And when people still push back?"

"There's Resistance Management for that," Susan said. "You identify the sources, listen, and address what's real. Sometimes resistance comes from smart people seeing real risks."

Megan leaned forward. "But once progress starts, you build momentum. Celebrate wins early, make success visible, and let that energy carry the organization forward."

Susan capped the marker. "Strategy isn't just what you plan. It's what people believe enough to execute."

## Common Execution Failures

Susan looked over the list on the slide and sighed. "This," she said, "is where even good strategies tend to die."

Jim frowned. "You mean during execution?"

"Yes," Susan replied. "The first killer is Over-commitment. Everyone wants to do everything at once, but bandwidth isn't infinite. When you spread attention too thin, everything slows down, and nothing gets done well."

Megan nodded. "And then there's Under-resourcing. Leadership sets bold goals but forgets to match them with people, time, or money. Ambition without capacity is just wishful thinking."

"Right," Susan said. "Next up, Poor Coordination. Teams chase their own goals without realizing how interconnected everything is. Strategy unravels when execution silos start pulling in different directions."

Jim leaned back. "So even if the plan's good, accountability still matters?"

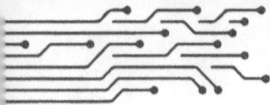

Susan pointed at him. "Yes. Weak accountability kills momentum. If nobody owns outcomes, then deadlines become suggestions and results stay vague."

Megan added, "And the hardest one, Change Resistance. You can have the best plan in the world, but if the culture doesn't want to move, strategy stalls. That's why communication and reinforcement matter just as much as structure."

Susan smiled faintly. "Execution isn't about doing more. It's about doing what matters, with discipline, clarity, and follow-through."

## Execution Integration

Megan clicked to the next slide, showing five boxes connected by arrows. "Here's where strategy either becomes real," she said, "or disappears into PowerPoint."

Jim raised an eyebrow. "Integration?"

"Yes," Megan replied. "Execution must live inside the organization's systems, not sit in a binder. Start with Budget Integration. If the money doesn't follow the strategy, nothing else will. Budgets are where priorities get real."

Susan added, "Then comes Performance Management. People should see how their daily work connects to strategic goals. When performance reviews reflect strategy, alignment follows naturally."

"Structure matters too," Megan continued. "If reporting lines or decision rights block collaboration, execution slows down. That's where Organizational Structure either helps or hurts you."

Jim nodded slowly. "And the data side?"

"Information Systems," Megan said. "You need dashboards, progress trackers, and communication tools that keep everyone synchronized. Otherwise, execution turns into guesswork."

Susan leaned forward. "Finally, Governance Processes. Someone must watch the watchers. Regular oversight keeps decisions honest and the strategy moving."

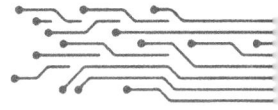

Megan smiled. "In short, integration turns strategy from intention into infrastructure."

## Goals

Moving forward, Susan began discussing goals, noting that they represent the strategic targets that define success for major organizational initiatives.

In the Vairos framework, goals serve as the North Star for execution, providing clarity, alignment, and accountability for strategic progress.

### The Strategic Role of Goals

When it came to the strategic role of goals, Megan was sketching something on a napkin. It was a mountain with three dots and a flag at the top.

"This," she said, "is what a goal really is. Not just a number or KPI, but a marker that says we're climbing toward something real."

Jim raised an eyebrow. "So... not just quarterly targets?"

"Targets are fine," she said, "but goals connect the dream to the day-to-day. They take the vision, the big 'why,' and turn it into something people can actually chase."

Susan jumped in. "And they've got to be sharp. If you can't measure it, you can't manage it. If there's no timeline, there's no urgency. People need to know what success looks like and when it's due."

Megan smiled. "But don't forget the human part. Goals should light a fire, not just fill a spreadsheet. A good one makes people want to win, not just check a box."

Jim looked at the napkin again. "So, it's not about climbing faster, it's about climbing with purpose."

"Yes," she said. "That's how goals turn strategy into movement."

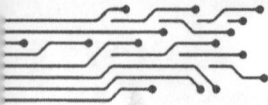

## Strategic vs. Operational Goals

Susan flipped through the deck of slides and sighed. "This is where most organizations trip up," she said. "They confuse running the business with changing the business."

Jim frowned. "Aren't both important?"

"Of course," Megan said. "But not equally. Strategic goals push you forward, such as market expansion, new capabilities, and new customers. That's transformation. Operational goals, on the other hand, keep the machine running, cost controls, processing efficiently and effectively with high quality consistently."

Susan leaned forward. "Think of it this way. Strategic goals build tomorrow's advantage; operational goals protect today's stability. If all your energy goes into operations, you'll be great at maintaining the status quo, until the market moves without you." Jim nodded slowly. "So, strategy is about deciding where to stretch, not just where to sustain."

"Yes," Megan replied. "Keep both but put your weight behind the ones that actually move the needle."

## Goal Development Process

Susan stood by the whiteboard with a marker in hand and said, "Before we lock in any goals, we need to make sure they fit. Alignment isn't just a buzzword; it's how you keep strategy from turning into chaos."

Jim tilted his head. "You mean making sure each goal connects to the mission?"

Circling the word alignment, Susan said, "If it doesn't move the organization toward its strategic direction, then it's really meaningless."

Megan added, "And don't forget reality. Ambition is great, but the market doesn't care about your enthusiasm. We need to validate every goal against what's happening out there, from competitors, demand to timing."

Susan nodded. "Then check your capabilities. Do we have the skills, systems, and people to pull it off? If not, build them or adjust the goal."

Jim smiled. "And I assume we need to be honest about resources and timelines, too."

"That's right," Megan said. "Dream big, but plan within reach. Unrealistic deadlines don't inspire; they destroy momentum."

## SMART Goal Framework

Susan clicked to the next slide. "Let's talk about how to make goals that actually work," she said. "Everyone loves grand statements, but if they're fuzzy, they're useless."

Jim laughed. "You mean like 'be the best in the industry'?"

"Ha ha, something like that," Susan replied. "That's not a goal, it's a wish. A real goal is specific; you should know exactly what you're trying to accomplish."

Megan chimed in. "And measurable. If you can't track it, you can't improve it. Numbers don't lie."

Susan agreed and said, "Then it becomes achievable. Stretch the team, sure, but not to the point where they burn out or stop believing it's possible."

Jim leaned forward and replied, "So it also must be relevant, right? No point chasing goals that don't support the big picture."

Adding to the conversation, Megan said, "And time bound. Every goal needs a finish line. Without deadlines, everything feels optional."

Susan smiled. "That's SMART goal design, not a corporate checklist, just common sense with structure."

## Goal Categorization and Portfolio

Susan drew a quick grid on the board. "Here's how we break goals into focus areas," she said. "Think of them as lenses. Each one captures a different part of performance."

Jim pointed at the first column. "Financial comes first, I assume?"

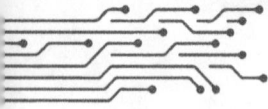

"Right," Susan said. "That's the scoreboard, revenue growth, profit improvement, cost control and return on investment. You can't fund ambition without financial health."

Megan added, "Then comes market goals, expanding share, winning new customers, maybe even entering new regions. That's where growth happens."

Susan moved to the next box. "Operational goals keep the engine running. These include efficiency, quality, capacity and process improvement. You can't scale chaos."

"And innovation?" Jim asked.

"That's your future insurance policy," Susan replied. "New products, new tech, new capabilities, everything that keeps you relevant."

Megan tapped the last square and said, "Don't forget stakeholders, customers, employees, regulators, and the community. If they're not with you, none of the other goals matter."

Jim nodded. "So, the art," he said, "is balancing all five without losing focus."

## Goal Measurement and Metrics

Susan tapped her pen against the conference table. "If you want to manage goals effectively," she said, "you need to measure the right things, not just what's easy to count."

Jim nodded. "You're talking about leading and lagging indicators again, aren't you?"

"Yeah," Susan said. "Leading indicators tell you what's about to happen, early signs of success or trouble. Lagging indicators tell you what will happen. You need both. One drives action, the other proves results."

Megan added, "And don't just wait for the big finish line. Track milestones along the way. They show whether you're moving in the right direction or drifting off course."

Susan turned toward the screen. "Benchmarking also matters.

Compare yourself to competitors and industry standards; it's the reality check leaders sometimes avoid."

Jim smiled. "And trend analysis ties it all together?"

"That's right," Susan said. "Trends tell the story. Whether you're improving, stalling, or slipping, and whether your strategy's working or just looking good on paper."

## Goal Ownership and Accountability

Susan looked up from her notes. "Ownership is where most strategies fall apart," she said. "Everyone loves the big announcement, but no one owns the follow-through."

Jim raised an eyebrow. "So where does ownership start?"

"With executive sponsorship," Susan said. "Senior leaders must champion the goals. Not just approve them, but more importantly own them. If the top tier doesn't care, then no one else will."

Megan leaned in. "Then it cascades. Each function or department takes responsibility for its piece. Finance, ops, marketing, everyone's got a part in the outcome."

Susan nodded. "And don't forget individual accountability. Every goal needs a name next to it. Not a committee, not a task force, a person. That's how you keep traction."

Jim crossed his arms. "But some goals need coordination across teams, right?"

Upon hearing this, Megan said, "That's where cross-functional work comes in. Silos kill strategy faster than bad ideas."

Susan smiled and said, "And finally, tie it to performance. When people see a clear link between achieving goals and earning recognition or rewards, accountability stops being a slogan and becomes part of the culture."

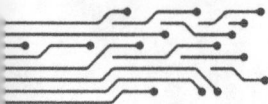

## Goal Communication and Alignment

Jim leaned back in his chair and said, "We've got a solid strategy, but how many people outside this room actually know what we're trying to achieve?"

Megan frowned. "Probably fewer than we'd like. Transparency still feels optional here."

Susan nodded. "And that's the problem. Organizational transparency isn't just about sharing updates; it's about alignment. When everyone understands the goals, execution stops being guesswork."

"Alright," Jim said, "so we start by making the goals visible?"

"Yes, that's right," Megan replied. "Post them, talk about them, and connect them to real work. When people see where the company's going, they naturally start to align their decisions."

Susan added, "And don't stop inside the building. Stakeholders, customers, investors, even partners, deserve to know what we're committed to. It builds confidence and credibility."

Jim pointed toward the screen. "But big goals don't mean much if departments can't translate them."

"Right," Megan said. "That's where department translation comes in. Each team should see how their objectives ladder up to the bigger picture."

"And individuals?" Jim asked. "That's the final link,"

Susan said. "Every employee should see how their daily work connects to those goals. It turns abstract targets into something personal."

Megan smiled. "And we keep the loop alive with progress reporting. We employ regular updates that show where we're winning, where we're behind, and what's next. It keeps everyone invested in the journey."

## Goal Adaptation and Evolution

Megan glanced at the dashboard on the screen. "Looks like we're ahead on revenue," she said, "but lagging in customer retention."

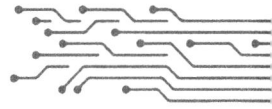

Jim nodded. "That's why performance monitoring matters. It's not just about checking boxes; it's about catching the story behind the numbers."

Susan leaned forward. "Exactly. You can't just set goals and walk away. Markets shift, competitors move, and what worked six months ago might not work now."

"So, course correction?" Jim asked.

"Adjust without losing direction. It's not admitting failure; rather, it's being smart enough to steer before you drift off course," Susan replied.

Megan added, "And when we make those adjustments, we should capture what we've learned. Every missed target, every surprise win, that's data for the next round of goal setting."

Jim pointed to the notes. "What about feedback? We've got customers, partners, and employees with front-line insights."

"Absolutely," Susan said. "Stakeholder feedback keeps us grounded. It's easy to get trapped in internal metrics, but the people we serve often spot the shifts before we do."

Megan smiled. "That's the whole point of market responsiveness: keeping goals flexible enough to move with the environment. If we can adapt fast, we stay relevant. If not, we end up reacting when it's too late."

## Common Goal-Setting Mistakes

Jim frowned at the whiteboard. "We had five goals last quarter," he said. "Three of them didn't even make sense when we looked back. "

Susan chuckled. "Classic vague objectives. If you can't measure it, you can't manage it. 'Improve engagement' doesn't mean anything unless you define what engagement looks like."

Megan added, "And sometimes we set great goals, but forget the resources. You can't expect a team of three to hit enterprise-level targets."

Jim said. "That's what happened with the analytics rollout. Ambition

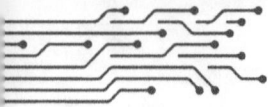

wasn't the issue, capacity was."

Susan pointed her pen at him. "That ties into timeline realism, too. Deadlines must reflect what's possible, not what looks impressive on paper."

Megan leaned back. "And we can't ignore the stakeholder angle. A goal that pleases leadership but frustrates customers or staff isn't a win, it's a warning."

Susan said. "And the last trap? Static thinking. We treat goals like contracts carved in stone when they should be living documents."

Jim smiled. "So, in short, be clear, be realistic, stay aligned, and keep evolving,"

Susan continued, "Otherwise, even the best strategy dies under the weight of its own goals."

## Goal Integration with Strategy

Susan drew a circle on the board labeled "Goals." "This," she said, "isn't just a target list. It's the engine for strategy."

Jim tilted his head. "You mean we build strategy from the goals, not the other way around?"

"Yes," Susan replied. "Strong goals refine direction. They show what really matters and force clarity about priorities."

Adding to the conversation, Megan said, "And once you know what matters, that's where resource allocation comes in. You don't spread your budget like peanut butter; you invest where the goals demand it."

Jim nodded. "So, goals drive where the money and people go."

"Right," said Susan. "And they tie into performance management too. Every individual and team should be able to see how their results move a strategic goal forward."

Megan added, "It's also about risk, I believe. Every goal carries exposure, financial, operational, or reputational. Tracking that helps

you see which ambitions are worth the gamble."

"Don't forget innovation," Susan said. "Goals should push the organization to build new capabilities, not just optimize old ones."

Jim smiled softly, "I see. So, the takeaway is, goals aren't a checklist. They're the strategic steering wheel."

## Goal Success Factors

Jim leaned back in his chair. "So, what really separates goals that succeed from the ones that die quietly in a slide deck?"

Susan didn't miss a beat before she answered. "Leadership commitment. If leaders aren't visibly behind the goals, or talking about them, funding them, and protecting them, then they won't stick."

Megan nodded. "Yeah, and commitment without resources is just pointless. You can't expect teams to hit targets with half the tools or budget they need."

Jim smiled. "So basically, no more champagne goals on a beer budget."

Susan chuckled and then continued, "Then there's alignment. Every system, performance reviews, reporting, and even incentives, must reinforce those goals. Otherwise, people end up chasing mixed signals."

Megan added, "And you need a culture that values performance. If accountability feels like punishment instead of pride, people will play it safe instead of pushing forward."

Susan capped her marker. "And don't forget to improve. Goal setting isn't a one-and-done process. The best organizations constantly refine how they define, track, and achieve their goals."

Jim nodded slowly. "So we need commitment, resources, alignment, culture, and improvement,"

Susan nodded. "When those five line up, goals stop being aspirations and start becoming results."

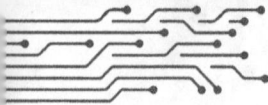

## Objectives

After covering the Goals section of the Varios framework, the three of them met a week later, and Susan began discussing objectives. She mentioned that they represent the specific milestones and deliverables that indicate progress toward strategic goals.

While goals define ultimate success, objectives break the journey into manageable components that maintain momentum and enable course correction throughout strategic execution.

### Objectives as Strategic Milestones

Megan leaned forward, tapping her pen against the table. "So, if goals tell us where we're going, objectives tell us how we'll know we're actually getting there, right?"

"Exactly," Susan said. "Objectives are the checkpoints, the milestones that keep us from drifting off course."

Jim nodded. "So, they're like mile markers on a highway. You hit one, you know you're still moving in the right direction."

"Right," Megan said. "Each one needs a clear result, something measurable. Not 'make progress,' but 'launch product beta by June.' That's what keeps teams motivated and accountable."

Susan added, "And timing matters. Objectives without deadlines are just wishes. You need urgency, or they'll always get pushed to next quarter."

Jim smiled. "And they should stack logically, right? One leading to the next, all the way to the final goal."

Susan pointed at him. "Exactly. Sequential logic. Each objective should make the next one easier, or at least more informed. That's how you build momentum."

Megan leaned back. "So, objective should be measurable, time-bound, sequential, and milestone-based. Always remember that objectives aren't just paperwork; they're proof that strategy is alive and moving."

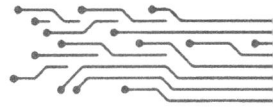

## The Goal-Objective Relationship

Jim looked at the whiteboard. "So, goals are the big targets, right? The end results we want?"

"Close," Susan said, picking up a marker. "Goals define what success looks like at the strategic level. They're directional, like a headline for what we're trying to achieve."

Pro-tip – Ask participants to create a front page for the local newspaper complete with a headline for each goal and a short story to go with the headlines

Megan added, "And objectives are the details underneath. They break that headline into steps you can manage and measure."

Susan drew two columns. "Allow me to give you an example here. Let's say our goal is to achieve a 10% market share in cardiovascular services in Region North by Q4 next year."

Jim nodded. "Alright, so what would the objectives look like?"

Susan began writing,

"Open two new cardiology clinics in Region North by Q2.

Recruit and onboard three new cardiologists by the end of Q3.

Establish referral partnerships with ten primary care practices by mid-year.

Achieve 50 cardiovascular procedures per month by Q3", she concluded.

Megan smiled. "Each one's a milestone that moves the needle toward that 10% target."

"Yes, that's right," Susan said. "Goals tell you where you're headed. Objectives make sure you get there."

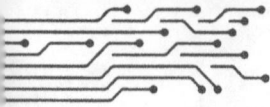

## Objective Development Process

Jim leaned back in his chair. "Okay, we've got our goals. But how do we turn them into something we can execute?"

"Through decomposition," Susan said, standing by the board. "You take a big strategic goal and break it down into the specific achievements needed to make it real. Think of it as unpacking success."

Megan nodded. "And you don't just list them randomly. You map out the timeline; which objectives come first, what depends on what, and how progress builds over time."

Jim frowned. "Similar to sequencing milestones?"

"Yes," Susan said. "Some objectives have dependencies. You can't open a new branch until you've hired the right team. You must understand how each piece connects."

Megan added, "And don't forget resources. Every objective needs people, time, and budget. It's pointless to set milestones you can't fund."

Susan smiled. "And finally, risk. Every objective has potential failure points. Spot them early on, plan your mitigations, and you'll keep the whole strategy from derailing halfway through execution."

## Types of Strategic Objectives

Susan drew a quick grid on the whiteboard. "Not all objectives are created equal," she said. "They serve different strategic purposes, capability, market, financial, and operational. Miss one category, and the whole strategy wobbles."

Jim gestured toward the top of the grid. "Let's start with capability. That's like building muscles before the marathon, right?"

Susan replied, "You hire specific talent, implement systems, and develop new processes. You have to complete all the behind-the-scenes work that lets the organization actually do what the strategy demands."

Providing further insight, Megan said, "Then you move to market

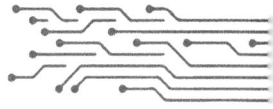

objectives. That's where you prove the strategy works in the real world: customer acquisition, market penetration, and brand recognition. Basically, you try to find out if customers notice and care?"

Jim pointed to the next section. "And the financial side ties it all together?"

"Right," said Susan. "Revenue, cost reduction, and profitability; they're the scorecard. Financial objectives tell you whether the strategy's sustainable or just a nice story."

Megan nodded. "And don't forget the operational layer, infrastructure, quality, efficiency, and compliance. It's not glamorous, but it's what keeps execution steady."

Susan capped her marker. "Each type plays a different role, but together, they're what turn vision into something measurable and real."

## Objective Measurement and Tracking

Susan leaned back in her chair. "Here's where most objectives fall apart," she said. "People say they've 'made progress,' but no one can define what that actually means."

Jim frowned and responded, "So how do you fix that?"

"Start with completion criteria," Susan replied. "Spell out what done looks like. No wiggle room. Everyone should agree when an objective has been achieved."

Megan nodded. "Also, between 'start' and 'done,' you need progress indicators and real metrics that show movement. Not feelings. Plain and crystal clear numbers."

Jim smirked. "And quality? That's the part everyone skips, right?"

"You can finish something and still fail strategically if it doesn't meet the quality standards tied to the bigger plan," said Susan.

Hearing this, Jim added, "That's why you monitor timelines and resource use too. Deadlines keep urgency alive, and tracking resources keeps ambition realistic."

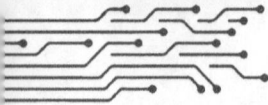

Susan smiled and replied, "If you don't define success before you start, you'll always think you're getting there, even when you're not."

## Objective Management Process

Jim scrolled through the project dashboard, frowning. "We've got fifteen objectives marked as 'in progress,' but half haven't moved in weeks," he said. "What's going on here?"

"Monthly reviews," Susan replied simply. "They're not just for reporting, they're for accountability. You sit down with the people responsible, look at what's moving, what's stuck, and why."

Megan gestured toward the screen. "That's where dashboards help. You can see everything succinctly: what's green, what's yellow, and what's bleeding red. No excuses and definitely, no surprises."

Jim sighed. "And when we find an obstacle?"

"You don't just note it," Susan said. "You fix it. Name the problem, assign ownership, and clear the path. Barriers don't disappear on their own."

Megan leaned forward. "And if the problem's bigger, then from resources, priorities to timing—you reallocate. Move people, money, or deadlines. The plan doesn't fail because of rigidity; it fails because no one adjusts."

Susan nodded. "Exactly. The point of reviews isn't to defend the plan, rather it's to keep it alive."

## Objective Ownership and Accountability

Megan leaned back in her chair. "Let's be honest," she said. "Half of our missed objectives aren't because people didn't care, they just didn't know who actually owned them."

Jim nodded. "So, we start with a clear assignment?"

"Exactly," Susan said. "Every objective needs a name next to it. One person, not a group. Ownership gets fuzzy the moment it's shared."

"And what if that person doesn't have enough authority to make things happen?" Jim asked.

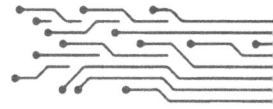

"Then that's leadership's fault," Megan said flatly. "You can't hold someone accountable for outcomes they don't control. Authority and responsibility must match."

Susan added, "Also, give them access to what they need, which mainly includes budget, data, and people. You can't set someone up to deliver and then starve them of resources."

Jim smiled. "So, performance reviews tie into this?"

"They have to," Susan said. "If people see that achieving objectives directly affects their evaluation and rewards, accountability becomes real."

Megan finished her coffee. "And when things stall, don't let it linger. There should be a clear escalation path and someone who can quickly unblock issues. Quickly end the waiting, no lost momentum."

## Cross-Functional Objectives

Jim glanced at the project chart on the wall. "These objectives touch half the company," he said. "How do we stop it from turning into chaos?"

"Coordination," Susan replied. "You can't run cross-functional work on goodwill alone. You need structure and clear processes for who does what, when, and how."

Megan added, "And consistent communication. Weekly updates, shared dashboards, open channels. If marketing changes a timeline, ops should know before it causes a bottleneck."

Jim frowned. "But what about when priorities clash? Finance says one thing, product says another."

"That's where conflict resolution comes in," Susan said. "You don't let issues fester. You must define upfront how decisions get made when there's a disagreement and who has final say."

Megan leaned forward. "And be realistic about resources. Shared teams mean shared budgets and shared time. Put it in writing so everyone knows the trade-offs."

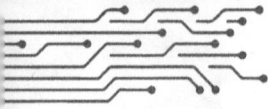

Susan smiled. "In the end, joint accountability is what keeps it honest. Success or failure belongs to everyone involved, not just the loudest department in the room."

## Objective Adaptation and Learning

The three of them then started talking about how to adapt and learn objectives. Leaning back in his chair, Jim said, "So once the objectives are set, do we just let them run their natural course?"

"Uhm, no, it's not that simple," said Susan. "You still need to check progress regularly and must always remember that an objective can still get stale pretty fast if you don't keep a check on whether they're still relevant and realistic."

Megan, nodding in agreement, said, "And when the environment changes, and it will sooner or later, you adjust, right? You don't throw out the strategy, but you do steer it in the right direction?

Hearing this, Jim said. "And what about the lessons we pick up along the way?"

"That's where the real gold is...," Susan said, "Capture what worked, what didn't, and why. Share it and always remember that the struggle one team experiences can be the success of the other."

Megan smiled. "That's how continuous improvement happens. Not through big slogans, but through small, honest conversations about what's working and what's not."

"To this", Jim said, "Yeah, I see the full picture now, that's how continuous improvement happens. It's not through big slogans, but through small, honest conversations about what's working and what's not. In a nutshell, that's where the real details are."

## Common Objective Management Mistakes

Sometime later, the three of them were reviewing a project that had fallen behind schedule. Jim sighed and said, "Looks like we've got another case of overly ambitious timelines. We keep setting deadlines that overlook how much we can realistically handle."

During the conversation, the three of them brought up a project that had fallen behind schedule.

Jim sighed and said, "Looks like we had another case of overly ambitious timelines. We kept setting deadlines that ignore how much we can realistically handle."

Susan nodded and said, "Yes, it's like we keep assuming people can stretch endlessly. Ambition's fine, but without capacity, it's just pressure disguised as progress."

Megan added, "And sometimes, even when we hit those deadlines, no one's quite sure what 'done' means. That's what happens when objectives don't have clear completion criteria — success becomes subjective."

"True," said Jim. "And when we underestimate the resources, time or money needed, or even just people, the whole thing collapses."

Susan leaned forward. "That's not all. Half the time, one team's working on something that doesn't sync with what another's doing. Poor coordination kills momentum."

Megan looked thoughtful. "And once the objectives are set, we treat them like stone tablets. But objectives should adapt. Conditions change, and management needs to as well."

Jim smiled. "So basically, our problem isn't ambition, it's rigidity, unclear goals, and not giving our people what they need."

To this, Susan replied, "It's not about setting fewer objectives. It's about setting smarter, flexible ones that have a chance to succeed."

## Technology and Objective Management

The conversation shifted toward tools. Jim looked at the screen and said, "So, how do we actually keep track of all these objectives without drowning in spreadsheets?" Susan smiled. "That's where project management tools come in. They're not just about task lists; they track dependencies, progress, and help everyone see how their work connects."

Megan chimed in, "And those dashboard systems you showed last week? They're a game changer. Real-time visuals of where we stand, no more guessing if we're behind or ahead." Jim nodded. "Makes sense. And for cross-team stuff, I guess collaboration platforms are the glue, right? Keeps everyone talking instead of working in silos."

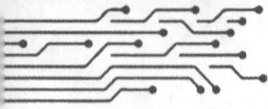

Hearing this, Susan said, "Add automated reporting to that mix, and you don't have to chase updates every Friday. The system just tells you what's on track and what's slipping." Megan leaned back, thinking. "And if we integrate analytics into all this, we're not just tracking progress, we're actually seeing how each objective contributes to the bigger strategy."

"Right," Jim said. "So, it's not about adding more tools. It's about connecting the right ones, so the strategy stays alive in real time."

## Objective Success Factors

The group sat around the whiteboard, now crowded with arrows and sticky notes. Jim tapped one with his pen and said, "Alright, we've got plenty of objectives written down, but how do we make sure they actually work?" Susan smiled. "It starts with clarity. Everyone involved should know exactly what the objective means and what success looks like. Ambiguity is what kills momentum."

Megan added, "And timelines. If they're not realistic, you're setting people up to fail. We must account for capacity and the unexpected, not just wishful thinking." Jim nodded thoughtfully. "So, it's about balance, not rushing, but not dragging either."

Hearing this, Susan said, "And don't forget resources. An objective without proper support is just a hope, not a plan." Megan leaned forward. "Ownership matters too. You need people who not only understand the objective but have the authority to move it forward."

"And that's where monitoring comes in," Jim said. "If we don't check in regularly, small issues can turn into big ones before anyone notices." Susan grinned. "You've got it. Clarity, realism, resources, ownership, and monitoring, get that right, and objectives stop being words on a slide and start becoming results."

## Strategies

After that, the three of them started talking about strategies, and Susan mentioned that strategies represent the fundamental approaches that organizations use to achieve competitive advantage and strategic objectives. In the Vairos framework, strategies define how organizations will compete, create value, and position themselves for long-term success.

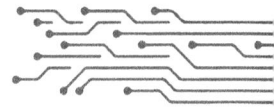

## Strategic Intent and Direction

The team gathered around the strategy board, the words Where, How, and What written in bold at the top.

"So," Jim began, pointing to the first line, "this part, where will we play? That's basically asking us to choose our battlefield, right?"

"Exactly," said Susan. "It's about focus. You can't win everywhere. You need to decide which markets, which customer segments, and which regions you're truly going to compete in."

Megan nodded. "And once you know that the next question is: How will we win? What's the edge we have that others don't? Is it our technology, our service model, our relationships, or something entirely different?"

Jim leaned back. "So, the second one's about differentiation. Making sure our strategy isn't just participation, but actual advantage."

"Right," Susan replied. "Then comes the third: What capabilities must we build? That's where reality meets ambition. You can't execute strategy without the right people, systems, and processes in place."

Megan smiled. "It's like building a bridge knowing where you're going, how you'll cross, and what materials you need to make it possible."

Jim chuckled. "So, strategy isn't just direction, it's discipline. Choosing, competing, and building for what really matters."

## Strategy as Guiding Decisions

Susan clicked to the next slide, a bold heading reading: "Strategic Decisions." She turned to Jim and Megan and said, "One of the biggest misconceptions in business is that strategy is a static document, something you write once, stick in a binder, and forget until the next planning cycle."

Jim laughed under his breath. "Sounds like every strategy I've ever seen."

"Exactly," Susan replied. "But the real strategy is alive. It's a continuous set of guiding decisions. These decisions determine where we invest

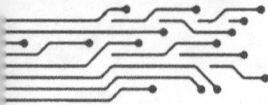

our resources, which capabilities we strengthen, and how we position ourselves to win. Most importantly, they keep the organization aligned while still allowing room for tactical flexibility."

Megan leaned forward. "So, you're saying strategy isn't about writing... It's about choosing?"

That's right," Susan said. "Strategy is a choice. And those choices fall into five major categories." She pointed to the list on the screen:

- **Market Focus:** Which customers and markets deserve priority attention and investment

- **Competitive Approach:** How we differentiate and create true competitive advantage with a moat around our business

- **Capability Investment:** Where we build strengths no one else can easily replicate

- **Partnership Strategy:** How we leverage external relationships to accelerate success

- **Innovation Direction:** Where we invest in new capabilities and market opportunities

Jim nodded slowly, taking notes. "So, these decisions are the links that keep everything connected?"

"Yes they are," Susan responded. "When made well, they create coherence. Every department, every initiative; decisions help usher all in the same direction."

Megan tapped her pen thoughtfully. "And if those decisions are bad?"

Susan smiled. "Then you have chaos. And chaos," she added, "is the most expensive strategy of all."

## Core Strategy Types

Susan advanced to the next slide. "Now that we've established strategy as a set of guiding decisions," she said, "let's look at the main strategic approaches organizations typically take to win in the market."

Jim raised an eyebrow. "Different ways to win. I like the sound of that."

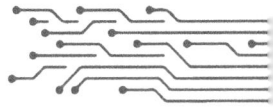

"There are five primary strategic paths," Susan continued, "and the trick is choosing the one that aligns with who you are and what you're capable of. Choosing the wrong one leads to strategic confusion." She pointed to the screen as she walked them through each approach.

"First, we have Differentiation. This is all about creating unique value, something that makes customers pick you and only you. It could be exceptional quality, breakthrough innovation, or a superior customer experience. Whatever it is, the goal is to stand out in a way competitors can't easily copy."

Megan nodded. "That sounds very consumer focused."

"It is," Susan replied. "The second approach is Cost Leadership. This doesn't mean being cheap; it means being efficient. Organizations that master this, run so smoothly that they can offer great prices while still protecting their margins."

Jim leaned back in his chair, deeply engrossed in the discussion. "Hospitals rarely do well on that one", he said.

"That's because healthcare has historically been built for professionals, not consumers," Susan said. "But that's changing fast." She clicked again. "The third approach is what we call Focus Strategy. This is where you go deep in a specific market segment or geography. You choose a niche, and you own it."

Megan smiled. "Specialists outperform generalists. That applies to business, too."

"Exactly," Susan agreed. "The fourth is Innovation Strategy. Here, competitive advantage comes from constantly developing new services, new technologies, or even entirely new business models. Innovation becomes a habit, not a one-time spark."

Finally, she brought up the fifth approach. "And then we have a Partnership Strategy. In today's interconnected world, no organization has every capability it needs internally. So, this strategy leverages alliances, joint ventures, and ecosystem relationships to expand reach and accelerate capability."

Jim folded his arms thoughtfully. "Five different paths to victory… and I'm guessing we have to pick one?"

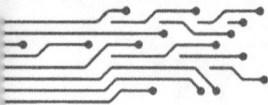

Susan smiled. "You pick a primary direction, yes. But the real power lies in how you combine these approaches without losing clarity. Strategy should give focus, not confusion."

"So," Megan asked, "how do we decide which path is right for Mega Health?"

"That," Susan replied, "is exactly what the Vairos framework will help us uncover."

## Strategy Development Process

Susan clicked on the next slide titled: "How Strategic Decisions Are made."

She looked at both Jim and Megan before speaking. "Now that we've covered the kinds of strategic choices organizations make," she said, "let's talk about how those decisions come together. Great strategy isn't luck; it follows a disciplined process."

Jim perked up, looking deeply interested. "Discipline is good. We need more of that around here."

Susan smiled. "Then you're going to really like this part. The decision-making workflow in Vairos has five key components." She pointed to the screen as each step appeared. "First, Strategic Analysis. This is where we assess reality: our strengths, weaknesses, market dynamics, and competitive forces. Basically, we're talking about everything that defines both our opportunities and our constraints here. We don't decide anything until we truly understand where we stand."

Megan nodded. "So, this prevents decisions based on assumptions."

"Exactly," Susan replied. "Next comes the Alternative Generation. We don't jump to the first idea that sounds good; we explore multiple pathways to success. Different market positions, different competitive angles and different levels of ambition."

She moved to the third step. "Then we have Evaluation and Selection. Here, we test each option against our capabilities, market conditions, and stakeholder needs. We ask: Can we execute this? Will it move us forward? Can we sustain it?"

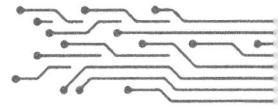

Jim tapped his pen. "You mean no more strategies built on wishful thinking?"

"Correct," Susan said, suppressing a laugh. "Fourth, Integration and Alignment. A strategy isn't a series of independent ideas. Everything must work coherently and support the mission and vision; otherwise, the organization pulls itself apart."

Megan raised her hand. "So, we avoid the trap of everyone working hard but not necessarily working together."

"You nailed it," Susan said. "And finally, Implementation Planning. Because a strategy is worthless until it becomes action. We will take those decisions and translate them into goals, objectives, and tactical initiatives that people can execute."

Jim leaned back in his chair with a satisfied nod. "This is the most practical strategic model I've seen. No fluff. Just clarity and execution."

"That's the point," Susan replied. "The Vairos framework doesn't just help us make smart choices; it ensures those choices turn into measurable success."

Megan smiled. "Then let's keep going. I want to know what comes next."

"And you will," Susan said, flipping to the next slide. "Because now we dive into how we make sure the organization stays aligned as we execute."

## Competitive Strategy Framework

Susan clicked to the next slide. "To make smart choices, we first need to understand the playing field. That's where Industry Analysis comes in. We look at the competitors, the market structure, the forces that shape profitability, everything that determines what success will require."

Jim nodded. "So not just who we're up against, but what drives the whole ecosystem."

"Exactly," Susan said. "Once we know that, we move into Competitive Positioning. We choose how Mega Health shows up in the market.

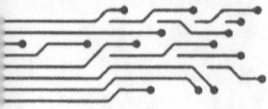

What makes us the better option? What strengths do we lean into?"

Megan folded her arms as she considered the slide. "So, it's a conscious decision about how we want customers to see us. Not just hoping they figure it out."

"Right," Susan said. "Positioning is intentional. Then we dive deeper with Value Chain Analysis. We break down every activity that creates value from product development to customer support and figure out where we can differentiate and where we can capture more value."

Jim smirked. "Translation: find the places where we can win, and double down."

Susan laughed. "Well said. But winning once isn't enough. We need Sustainable Advantage, things competitors can't easily copy. Maybe it's proprietary tech, maybe unmatched expertise, maybe relationships that take years to build."

Megan leaned forward. "And once we have that, we protect it."

"Exactly," Susan replied. "Because markets change. Competitors react. So, we study Dynamic Competition, we anticipate responses, prepare countermoves, and maintain our edge as conditions shift."

Jim looked impressed. "So, strategy isn't a one-time decision. It's staying sharp, staying aware, and staying ahead."

Susan nodded with a confident smile. "That's the mindset that keeps Mega Health winning."

## Strategic Coherence and Integration

Susan advanced to the next slide. "Now, even the best strategy falls apart if the organization isn't aligned behind it. Effective strategy creates coherence; everything moves in the same direction."

Jim raised an eyebrow. "Meaning no teams running off doing their own passion projects?"

"Exactly," Susan said with a smile. "It starts with Resource Alignment. Our financial, human, and technology investments must back the strategy, not compete with it."

Megan nodded. "Show me a budget, and I'll show you what a company truly values."

"Perfectly put," Susan replied. "Next is Capability Integration. Our skills and competencies must reinforce the position we're claiming in the market. If we say we're the best in patient experience, our capabilities must prove it."

Jim tapped his pen. "So, no hollow promises."

"Right," Susan continued. "Then there's Cultural Consistency. Strategy dies fast if behaviors don't match it. We need values and everyday actions that support how we want to compete."

Megan smiled knowingly. "Culture is either the silent killer of strategy, or its greatest amplifier."

Susan clicked again. "Which brings us to System Coordination. Policies, processes, and structures must enable execution, not create friction."

Jim let out a light laugh. "No more approvals that require three committees and a blood sample."

"Exactly," Susan replied. "Finally, Stakeholder Alignment. We need customers, employees, partners, or rather everyone in our ecosystem, moving with us, because strategy is a team sport."

Megan leaned back and, with a pleased look, said, "So alignment isn't a bonus. It's the engine."

"And when we get this right," Susan concluded, "execution becomes smoother, faster, and far more effective."

## Strategy Communication and Embedding

Susan shifted to the next slide and said, "Alignment doesn't stop at systems and resources. It starts with the people at the top. Leadership Alignment means every leader understands the strategic direction and is fully committed to it."

Jim chuckled. "So, no more 'secretly running my own strategy in my department' attitude."

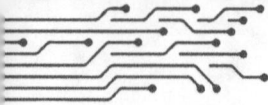

"Exactly," Susan said. "When leaders disagree privately, the organization gets mixed messages publicly. This is why Organizational Communication is critical, translating strategy into clear and relatable language that everyone can understand."

Megan nodded with a thoughtful expression. "People can't execute what they don't grasp."

"Right," Susan continued. "Then we build Department Integration. Every function—operations, marketing, and clinical—must connect its plans to the enterprise strategy. No silos. No misalignment."

Jim tapped his notepad. "The heartbeat of execution."

"And it gets even more personal," Susan said. "Individual Connection means every employee sees how their day-to-day work contributes to strategic success. When people understand their impact, engagement skyrockets."

Megan smiled. "That's how you turn a workforce into a movement."

"Exactly," Susan said approvingly. "And when we embed these ideas deeply, or when strategic thinking becomes part of how we make decisions and behave daily, Cultural Embedding occurs. Strategy becomes who we are, not just what we say."

Jim looked satisfied. "Alignment isn't just communication, it's transformation."

"And it's what turns great plans into great performance," Susan said, advancing to the next slide.

## Strategic Adaptation and Evolution

Continuing with her presentation, Susan said, "Now we will reflect on the discipline of staying sharp over time. Strategy isn't 'set it and forget it.' It requires constant awareness. We start with Environmental Monitoring. This includes watching shifts in the market, regulations, technology, and customer expectations so we can adjust before threats hit us."

Jim nodded slowly. "So instead of reacting when it's too late, we move first."

"Exactly," Susan said. "Then we have Performance Evaluation. We regularly assess whether the strategy is working and whether we're improving our competitive position."

Megan glanced over at Jim. "Metrics. Accountability. No wishful thinking."

Susan smiled. "That's the spirit. Next is Learning Integration. Every strategic initiative teaches us something, good or bad. We take those lessons and refine the strategy instead of repeating mistakes."

Jim leaned back. "So, the strategy gets smarter as we go."

"Right," Susan continued. "And we must expect competitors to respond. With Competitive Response, we watch their moves, anticipate their counters, and adjust our strategy so we stay ahead rather than catch up."

Megan crossed her arms with a grin. "A proactive chess match."

"You could call it that," Susan agreed. "And finally, Innovation Integration. Strategy can't just protect what works today; it must open doors to what will work tomorrow. We link strategic direction with experiments, new technologies, new business models, and the things that create future advantage."

Jim exhaled, impressed. "So, strategy execution isn't maintenance. It's momentum."

Susan nodded with confidence and said, "And that momentum is what keeps MegaHealth resilient and unbeatable."

## Strategy Implementation Challenges

Susan moved to another slide. "Of course, strategy isn't sunshine and simplicity. We must navigate the real obstacles that get in the way. First up: Resource Constraints. We always want to do more than we realistically can, financially, technologically, and in terms of talent."

Jim sighed knowingly. "So, it's about discipline. Choosing what to do and what to delay."

"Exactly," Susan said. "Then there's Organizational Resistance. People

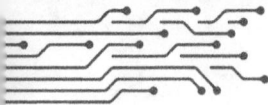

don't naturally love change. Some fear it. Some politically resist it. And if we don't manage that, the strategy stalls."

Megan gave a small laugh. "Culture has veto power if you ignore it."

"Right," Susan agreed. "And just like before, Competitive Response plays a role here too. When we make bold moves, competitors react.

We must anticipate how they'll try to neutralize us."

Jim leaned forward. "Strategy we execute, not strategy they can shut down."

"Precisely," Susan replied. "Then comes Market Evolution. The customer needs a shift. New expectations emerge. Strategy must adapt or it becomes outdated."

Megan nodded. "Standing still is the fastest way to lose relevance."

"And finally," Susan said, clicking the last bullet, "Execution Complexity. Turning strategy into real action introduces interdependencies, coordination challenges, and change management hurdles. It's messy, unless we manage it intentionally."

Jim smiled as it all cametogether. "So, strategy isn't difficult because the ideas are wrong; it's difficult because reality fights back."

"Yes, that's right," Susan said. "And that's why we plan not just for the strategy itself, but for everything that can challenge it along the way."

## Common Strategic Mistakes

Continuing with the presentation, Susan said, "There are also classic strategic mistakes we must avoid. The first is Strategy Proliferation, which means trying to do everything at once. When companies chase too many priorities, none of them win."

Jim, in agreement with Susan, replied, "Ah, yes, the 'we can be all things to all people' dream that turns into a nightmare."

"Exactly," Susan said, smiling. "Then we have Generic Positioning. Strategies that sound nice but don't set us apart. If customers can't tell the difference, we didn't choose a real position."

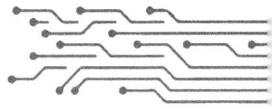

Megan nodded. "So, we either stand out... or we disappear."

"Right," Susan continued. "Another pitfall is Capability Mismatch. Picking a strategy that requires strengths we don't have and can't build quickly enough."

Jim raised an eyebrow. "Like claiming world-class digital experience without world-class digital talent."

"Perfect example," Susan replied. "Then there's Market Misunderstanding. Strategy built on faulty assumptions, wrong customers, wrong demand signals, and wrong trends. If we're wrong about the world, the world wins."

Megan smirked. "And the market never adjusts to make us feel better."

"It would be nice if it did," Susan laughed. "And finally: Implementation Neglect. Companies fall in love with big ideas but forget execution. Strategy without follow-through is just hope in a PowerPoint deck."

Jim pointed at the slide. "So, discipline, differentiation, realism, accuracy, and execution. Those are the guardrails."

"Yes," Susan said firmly. "Avoiding these mistakes isn't optional; it's how we keep strategy real, practical, and successful."

## Strategic Success Factors

Susan brought up the next slide. "So, what defines a strong, well-designed strategy? First, it provides clear direction. People know what to prioritize, and what to stop doing."

Jim nodded. "No guesswork. No mixed signals."

"Yes," Susan said. "Next is Distinctive Positioning. A good strategy makes us meaningfully different, not just another player in the market."

Megan smiled. "Differentiation is the price of admission to relevance."

"Well put," Susan continued. "Then comes Capability Alignment. A strong strategy leverages what we're already great at, while building the strengths we need for the future." Jim tapped his notebook. "It plays to our edge."

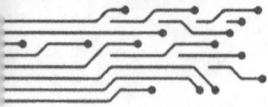

"Right," Susan replied. "And we can't ignore Market Relevance. Strategy must be grounded in real customer needs and emerging opportunities, not internal assumptions."

Megan leaned forward. "The market decides who wins. We adapt to them, not the other way around."

"Yes, that's right," Susan said. "And finally, Implementation Focus. Strategies don't stay in theory; they convert into specific initiatives that people can act on."

Jim gave a satisfied nod. "So great strategy equals clarity, differentiation, strength, relevance, and action."

Susan smiled. "That's the formula for competitive advantage and for MegaHealth's future success."

## Strategy and Innovation Integration

Susan switched to a fresh slide and said, "Now let's talk about how innovation fits into strategy, because growth doesn't come from standing still. First, we have an Innovation Strategy. It's a disciplined approach to developing new capabilities and exploring new markets, not random creativity."

Jim chuckled. "So, innovation isn't just 'cool ideas' day?"

"No, it's not," Susan smiled. "It's direction with purpose. Then we elevate that to Strategic Innovation, using innovation itself as a source of competitive advantage. It's how we create new positions in the market, rather than just defend old ones."

Megan nodded thoughtfully. "Innovation that changes the game, not just improves the score." Susan agreed and continued with the presentation. She said, "Next is Technology Integration. We deliberately use technology to strengthen our strategic position, better experiences, smarter operations, and faster decisions."

Jim leaned in. "Tech becomes the accelerator of the strategy."

"Well said," Susan replied. "Then we explore Business Model Innovation, finding new ways to deliver value and generate revenue. Sometimes the biggest breakthroughs aren't products, but the way we package,

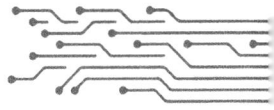

deliver, or monetize them."

Megan smiled. "That's how you turn disruption into opportunity."

"And finally," Susan said, pointing to the last bullet, "there's Partnership Innovation. We rethink collaboration, alliances, ecosystems, and co-development, to access capabilities and markets we can't reach alone."

Jim sat back, impressed. "So, innovation isn't a department. It's a strategic engine."

Susan nodded. "Exactly, and when we harness it correctly, we create the future instead of reacting to it."

## Tactics

After strategies, Susan moved on to tactics, which represent the specific initiatives, projects, campaigns, and actions that implement strategic approaches and achieve strategic objectives. While strategies define broad direction, tactics specify the detailed work that transforms strategic intentions into organizational reality.

### Tactics as Strategic Implementation

Susan moved to the next slide. "Now, let's talk about tactics, the practical bridge between strategy and day-to-day action. Strategy points us in the right direction. Tactics make the journey real."

Jim leaned in. "So, this is where planning finally becomes work?"

"Yes, you're spot on," Susan said. "Tactics translate those big strategic ideas into specific activities we can assign, schedule, and measure. They're the 'how.'"

She clicked to highlight the first point. "Specific Actions, each tactic is a clearly defined initiative with a precise scope and deliverables. No ambiguity."

Megan nodded. "If the description is fuzzy, the results will be too."

"Right," Susan continued. "And every tactic must have an Assigned

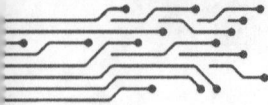

Responsibility. Someone must own it; not 'the organization,' but an accountable individual or team."

Jim smirked. "Ownership prevents the classic 'I thought YOU were doing it' conversation."

Susan smiled. "Then we set Defined Timelines, actual start and end dates, so progress is visible and urgency is real." Megan tapped her pen. "Deadlines are promises to the future."

"Exactly," Susan agreed. "Next: Resource Requirements. We determine what budget, talent, and technology are necessary before we begin."

Jim wasn't too keen on this and said, "Execution without resources equals frustration."

"Well said," Susan replied. "And finally, Measurable Outcomes. Tactics produce quantifiable results. We know if they worked, and how well."

Megan looked satisfied. "So, strategy is the map. Tactics are the roads. And execution is the drive."

"Perfect analogy," Susan said. "And Mega Health will master all three."

## The Strategic Execution Stack

Susan clicked forward and said, "Okay, now let's clear up a common confusion: Strategy, goals, objectives, and tactics are not the same thing. They stack together, each one more specific than the last."

Jim crossed his arms. "And when organizations mix those up, everything becomes chaotic."

"You're not wrong," Susan said. "The mix up between these is such a common occurrence in organizations. So I thought, why not break it down? I'll start with strategy. The latter is the broad approach; how we plan to win and gain a competitive advantage."

Megan jumped in. "The big choice. The direction."

"Yes," Susan agreed. "Then we set a Goal, a measurable outcome that shows whether the strategy is working."

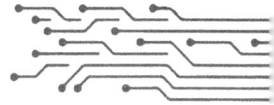

Jim nodded. "Not fluffy ambition. Actual proof and from there, we define Objectives, or specific milestones that move us toward the goal."

Megan smiled. "Steps on the path."

"Exactly. And finally, Tactics, the concrete initiatives that make those objectives happen day by day."

Jim leaned forward. "Give me a real example."

At Jim's request, Susan brought up a scenario on the slide:

- Strategy - Expand cardiovascular services in Region North

- Goal - Achieve 10% market share by Q4 next year

- Objective - Open two new clinics by Q2

- Tactic - Launch buildout project and hire a project manager by the end of the month

Jim pointed at the screen, impressed. "That's clean. I can see exactly how each step leads to results."

"That's the point," Susan said confidently. "When these four levels are aligned, execution becomes inevitable."

Megan smiled and said, "Clarity creates momentum."

"And momentum," Susan added, tapping the table lightly, "creates market leadership."

## Tactical Categories and Types

Susan moved to the next slide and said, "Now that we've clarified how tactics connect to strategy, let's look at the four major types of tactics organizations rely on. Each type has a distinct purpose, and the right balance determines whether strategy becomes reality."

Jim, rubbing his chin, replied, "So this is where execution actually happens."

"Yes," Susan said. "We'll start with Market Development Tactics; the

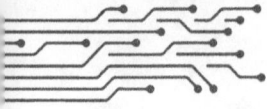

things we do to grow visibility and build customer relationships."

She listed examples on the screen:

- Customer acquisition campaigns
- Market research and analysis
- Brand development initiatives
- Distribution channel expansion

Megan nodded and said, "So these are about presence and perception."

Susan continued, "Next, we have Capability Building Tactics, which is about strengthening the organization internally so we can deliver on the strategy."

She highlighted another set of bullets:

- Talent recruitment and development
- Technology implementation
- Process improvement
- Training programs

Jim pointed. "That's investing in muscle before we try to lift more weight."

Susan smiled. "Perfect analogy. Now, Operational Excellence Tactics are all about making the machine run better."

- Efficiency improvement
- Quality enhancement
- Cost reduction
- Performance optimization

Megan leaned forward. "That sounds like where hospitals spend most of their time now, doing more with less."

"And often," Susan said, "they forget the Innovation Tactics, the ones

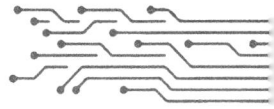

that unlock future opportunities." She flipped to the final list:

- Research and development projects
- New product or service creation
- Technology pilots
- Partnership development

Jim exhaled. "So, execution is like managing a portfolio, market growth, stronger capabilities, smoother operations, and future innovation."

"Yes," Susan replied. "Strategic success comes from choosing the right mix of these tactics, not just doing what's easy, familiar, or urgent."

Megan smiled. "I can already see how Vairos will help us balance those decisions."

"And that balance," Susan said, closing the slide, "is the difference between playing catch-up and leading the market."

## Tactical Planning Process

Susan brought the next slide on. "Turning strategy into action requires structure. We will break it down step-by-step."

"Walk us through it," Jim said. "First," Susan began, "Objective Decomposition. In this, we take each objective and identify exactly what must get done. Then, Initiative Design, every tactic gets a clear scope, timeline, and resource need."

Megan added, "So no ambiguity, everyone knows what success looks like."

"Right," Susan replied. "Which leads to Resource Allocation and Timeline Development, we commit budget, talent, and realistic schedules."

Jim pointed at the screen. "But someone must own it…"

"And they do," Susan said. "Responsibility Assignment makes one person accountable for each initiative. Finally, we set Success Metrics so we can measure progress, not just hope for it."

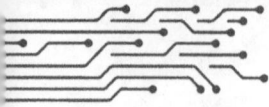

Megan smiled. "This is execution with discipline."

"And that," Susan concluded, "is how strategy becomes results."

## Tactical Project Management

Susan clicked to the next slide. "Execution isn't just starting projects, it's managing them with discipline,"

Jim nodded. "Meaning no surprises?"

"Exactly," Susan said. "We begin with Project Definition, clear scope, deliverables, and what success looks like. Then Timeline Management, which includes realistic schedules with room for the unexpected."

Megan leaned in. "And the resources?"

"Coordinated from the start," Susan replied. "Resource Coordination ensures the right people, money, and technology are in place. Next, Quality Assurance, execution must meet strategic standards, not just get completed."

Jim smiled. "And we track progress, right?"

"Every step," Susan said. "Progress Monitoring keeps us ahead of obstacles, and Issue Resolution removes them quickly, so momentum never stalls."

Megan sat back with a thoughtful grin. "This is how strategy stays alive."

"And how organizations," Susan finished, "turn intent into impact."

## Tactical Flexibility and Adaptation

Susan moved to a new slide. "Even with strong planning, execution must stay flexible. Strategy doesn't freeze, neither can tactics."

Jim raised an eyebrow. "So, we're talking agility?"

Susan replied. "Agile Execution means we adjust tactics when reality gives us new information. Then comes Performance Monitoring,

continuously checking if each tactic is helping the strategy."

Megan added, "And if it isn't... we change course?"

"Yes," Susan said. "Course Correction keeps us aligned with our objectives while still adapting. We also practice Learning Integration, capturing lessons as we execute."

Jim summarized, "So, the organization gets smarter while it moves."

"Precisely," Susan said. "Which leads to **Rapid Iteration,** test fast, refine fast, succeed faster." Megan nodded with approval. "Not just executing, evolving."

"And that," Susan said, "is how we stay ahead instead of catching up."

## Cross-Functional Tactical Coordination

Susan brought up the next slide. "Most tactics don't live in one department: they cross boundaries. That's where coordination becomes critical."

Jim nodded. "Hospitals are famous for silos."

"Which is why we start with Integration Requirements," Susan explained. "We map out which functions depend on each tactic, so no one is surprised later."

Megan pointed to the slide. "And all of that means more communication?"

"Exactly," Susan said. "Communication Protocols ensure regular updates between teams. We also manage Resource Sharing, because multiple initiatives often need the same people or tools."

Jim smirked. "And when teams fight over those resources?"

Susan smiled knowingly. "That's why we establish Conflict Resolution procedures before conflict shows up. Finally, we use Collaborative Planning when a tactic requires cross-functional execution from day one."

Megan exhaled. "No more tug-of-war. Just organized teamwork."

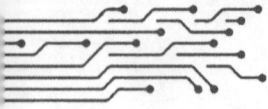

"And that," Susan concluded, "is how complex tactics succeed instead of colliding."

## Tactical Performance Management

Jim leaned forward as Susan switched to the next slide. "Tracking tactics sounds simple, but I'm guessing it's not?"

"Not simple," Susan confirmed, "but essential." She pointed to the screen. "We start with Weekly Reviews, quick check-ins to confirm what moved forward and what's stuck."

Megan raised a brow. "So, we're not waiting a month to find out something went wrong?"

"Exactly," Susan replied. "Then we use a centralized Dashboard, so everyone sees tactical status in one place, no guessing, and no hidden delays."

Jim tapped his notebook. "Deadlines still matter, though, right?"

Susan nodded. "That's why we monitor Milestone Tracking. Every major deliverable has a target date, and we track completion against that plan."

Megan glanced up. "And if we're overspending or stretching the team too thin?"

"We track Resource Utilization," Susan said. "It shows whether we're using our people and budget wisely or burning through them."

Jim smiled. "Okay, but how do we know the work is actually good?"

"Quality Assessment," Susan responded confidently. "We test whether the results match what the strategy requires, before we call anything 'done.'"

Megan leaned back, satisfied. "Weekly clarity, visible progress, smart resources, and quality checks."

"And that," Susan concluded, "is how we keep tactics on course, instead of letting them quietly drift."

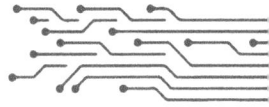

## Tactical Excellence through Technology, Discipline, and Learning

Susan brought up a new slide and said, "Tactics succeed when technology, discipline, and learning work together. Let's start with the tech side."

Jim nodded. "We've invested in software, but I'm not sure we're using it strategically."

"That's common," Susan said. "Project management systems keep every initiative visible, including progress, dependencies, and resources. Then collaboration tools and communication platforms keep teams connected and decisions flowing."

Megan added, "And dashboards?"

"Yes," Susan said. "Performance dashboards show real-time status, while resource management systems help balance capacity across multiple tactics."

Jim leaned forward. "And when things still go wrong?"

Susan smiled. "That's where discipline comes in. The biggest traps are scope creep, resource underestimation, and poor coordination. Add timeline optimism and quality shortcuts, and execution starts to unravel."

Megan frowned a little and said, "So how do we prevent that?"

"By focusing on the success factors," Susan replied. "Clear definition of every tactic, adequate resources, strong ownership, and regular monitoring. Most importantly, every tactic must stay linked to strategy; otherwise, we're just staying busy."

Jim nodded slowly. "And the learning piece?"

"Critical," Susan said. "We analyze performance, capture best practices, and drive process improvement. Then we spread that knowledge, transfer it, and build capability development so the organization gets stronger with each cycle."

Megan smiled. "So, technology gives visibility, discipline ensures focus, and learning drives evolution."

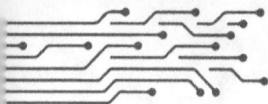

Susan nodded. "Exactly. Tactical excellence isn't luck, it's a system."

Susan advanced the slide. "Strategy becomes real through tactics; they're the actions that make direction tangible."

Jim nodded. "So tactical performance tells us if the strategy's working."

"Yes, you're right," Susan said. "Each tactic drives strategic progress, informs resource planning, and shapes timelines. If tactics drift, the whole strategy wobbles."

Megan leaned forward. "So, visibility matters."

"It does," Susan replied. "Project systems, dashboards, and collaboration tools keep us aligned; everyone sees progress, dependencies, and resource use in real-time,"

Jim smirked. "And when things go wrong?"

"Then we guard against the usual traps, scope creep, poor coordination, unrealistic timelines, and quality shortcuts," Susan said. "Success depends on clear definition, adequate support, and strong ownership."

Megan crossed her arms. "And learning?"

Susan smiled. "Always. We analyze results, share best practices, improve processes, and transfer that knowledge across teams. That's how we build capability and refine strategy."

Jim nodded. "So, tactics aren't just execution, they're evolution."

"Yes," Susan said. "When tech, discipline, and learning align, strategy doesn't just happen. It thrives."

## Policies

After that, all of them started talking about policies and stated that they serve as the strategic guardrails that enable autonomous execution while maintaining organizational coherence and compliance.

In the Vairos framework, policies are not bureaucratic constraints but

strategic enablers that provide clarity and consistency for decision-making throughout the organization.

## Policy as Strategic Infrastructure

Susan brought up the next slide. "We've talked about tactics and execution, now let's discuss the framework that keeps everything aligned: policy."

Jim frowned slightly. "Policies? Those usually slow things down."

"Only bad ones," Susan said with a smile. "Good policies act as strategic infrastructure; they define how we operate, so decisions stay consistent, even when leadership isn't in the room."

Megan nodded. "So, they're not about control, they're about clarity."

To this, Susan replied, "Policies reduce ambiguity, ensure compliance, and protect the brand. They also enable scale, we can grow without micromanaging, and most importantly, they support strategy by aligning daily actions with long-term priorities."

Jim leaned forward. "What kinds of policies are we talking about?"

Susan clicked through the categories. "First, Strategic Policies guide big decisions like investments, partnerships, innovation, and competitive response. Then, Operational Policies, quality standards, customer service expectations, vendor management, and performance criteria."

Hearing this, Megan added, "And compliance fits in there somewhere, I assume?"

"Right," Susan said. "Compliance Policies cover data privacy, financial reporting, safety, and ethics. Finally, we have Governance Policies, which outline how authority, communication, risk, and change are managed."

Jim tapped his pen thoughtfully. "So, policy isn't bureaucracy, it's structure."

"Exactly," Susan replied. "When policies are clear and connected to strategy, they don't slow us down; they make us faster, smarter, and safer."

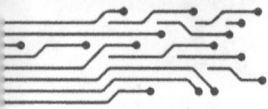

Megan smiled. "Then they really are the backbone of strategic execution."

Susan nodded. "That's the idea; policies turn strategy into consistent behavior across the organization."

## Building Smart, Adaptive Policies

Susan flipped to the next slide. "Policies aren't about red tape, rather they're strategic tools. If you do them right, they keep the organization aligned and agile."

Jim crossed his arms. "So, they're part of strategy, not just compliance?"

"Yes, you're right," Susan said. "We start with strategic alignment, gather stakeholder input, study best practices, assess risk, and design policies that can actually be implemented and improved over time."

Megan leaned forward. "And they tie directly to initiatives?"

"They have to," Susan replied. "We look at strategic dependencies, make sure we meet compliance needs, manage risk, and reinforce culture and performance standards that match our goals."

Jim nodded. "But what keeps policies from going stale?"

"Dynamic management," Susan said. "Regular reviews, feedback from users, and strong change management so updates roll out smoothly."

Megan smiled. "And communication?"

"Critical," Susan said. "Clear documentation, training, and decision tools help people apply policies daily. Plus, ongoing communication and feedback keep them relevant." Jim raised a brow. "And the usual mistakes?" Susan ticked them off. "Too complex, too vague, or focused only on compliance. Others forget to review or communicate."

"So, what makes them work?" Megan asked.

"Relevance, practicality, clarity, regular review, and cultural fit," Susan said. "When policies align with purpose, they don't slow us down; they make us sharper."

## Measure

All of them decided to end the meeting on the policy discussion. They met again a week later to discuss another essential component: Measure.

While talking about this, Susan said it transforms strategic execution into quantifiable insights through systematic performance tracking and analysis. Measurement provides the feedback loop that enables learning, accountability, and continuous improvement throughout the strategic planning cycle.

Susan dimmed the lights slightly as a new slide appeared and then said, "Measurement isn't just about tracking performance," she began. "In Vairos, it's how we learn, adapt, and prove value."

Jim leaned forward. "So, it's not just numbers on a dashboard?"

"Yes, that's right," Susan said. "It's about progress validation, course correction, accountability, learning, and stakeholder communication. Measurement turns strategy into a living feedback loop."

Megan glanced at the chart and said, "So what does that system look like in practice?"

Susan pointed to the screen. "We build a measurement architecture, dashboards for visibility, structured evaluations for effectiveness, and feedback mechanisms to capture lessons as we go."

Jim nodded. "And I assume not every measure is created equal."

"Right," Susan replied. "We measure at four levels, strategic, goal, objective, and activity, blending leading indicators that predict success with lagging ones that confirm results."

Megan smiled. "And those measurements feed back into planning?"

"They have to," Susan said. "Performance data shapes strategic choices, resource allocation, and implementation management. It even refines how we communicate progress to stakeholders." Jim sat back, impressed. "So, measurement isn't just about control, it's about intelligence."

Susan smiled. "Exactly. The Measure component makes strategy smarter every cycle. It ensures we don't just execute; we evolve."

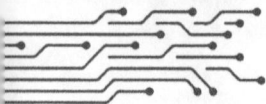

## Dashboard

Next, she discussed strategic cards and said they provide real-time visualization of organizational performance across all major strategic initiatives. In the Vairos framework, dashboards serve as the strategic cockpit, enabling progress, identifying issues, and informing resource-allocation decisions and course corrections.

Susan brought up a sleek dashboard on the screen. "This," she said, "is where strategy meets reality. Dashboards are our strategic cockpit; they turn mountains of data into something leaders can use."

Jim squinted at the display. "So, this shows everything happening across Mega Health in real time?"

"Exactly," Susan said. "They give real-time visibility, flag exceptions that need attention, and reveal patterns across operations. More than that, they support decision-making and accountability, everyone sees that we're winning and where we're drifting."

Megan tilted her head. "But one dashboard can't fit everyone's needs, right?"

"Right," Susan replied. "Hence, we design them by role. Executives get high-level strategic progress, managers see tactical performance, and individuals track their own objectives. The key is clarity and actionable insight, not information overload."

Jim pointed at a chart. "What about the mix of metrics?"

"A balanced perspective," Susan explained. "We combine financials, operations, customer metrics, and innovation indicators. For leadership, that includes goal progress, KPI trends, initiative status, resource use, market position, and risk alerts."

Megan smiled. "And for teams on the ground?"

"They see what matters most to their world," Susan said. "Project timelines, performance metrics, resource utilization, and even collaboration updates. It connects everyone from the boardroom to the front line."

Jim leaned back, impressed. "So, the dashboard doesn't just report, it drives behavior."

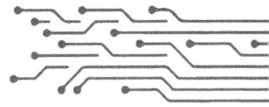

"Yes, you're on the right track here," Susan said with a grin. "When the data is clear, the direction is too. That's how we keep strategy alive, one dashboard at a time."

Susan tapped the screen. "Think of this dashboard as the organization's nervous system. It tells us what's working, what's slipping, and where we need to act."

Jim squinted at the visuals. "And it pulls data from everywhere?"

"Yes," Susan said. "Financials, operations, HR, customer metrics, all integrated in real time. You can even access it on your phone, so you're never out of the loop."

Megan pointed to a blinking alert. "So, we know immediately when something goes off track?"

"Yes," Susan replied. "Alerts flag exceptions, and the dashboard can be customized by role. Executives see strategy-level insights, managers get tactical progress, and individuals track their own objectives."

Jim leaned back. "And the metrics we track?"

"All the essentials," Susan said. "Financial health, customer satisfaction, operational efficiency, innovation, employee engagement, and market position. Together, they give a full picture of performance and risk."

Megan nodded. "How do we make sure people actually use this?"

"Through regular review cycles, exception analysis, trend monitoring, and turning insights into action plans," Susan explained. "But dashboards fail if you are overwhelmed with metrics, focus only on lagging indicators, or present data poorly. Stale information and missing context are killers too."

Jim smirked. "So, what separates a good dashboard from a bad one?"

Susan counted off on her fingers. "Strategic relevance, clarity, timeliness, actionable insights, and user adoption. If people don't look at it or act on it, it's useless."

Megan smiled. "So, it's more than just data; it's a decision engine."

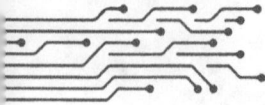

"Yep," Susan said. "When dashboards work right, they integrate planning, performance, risk, and communication. They don't just show strategy; they make it happen."

## Evaluate

Later in the week, Jim, Susan, and Megan met for another session to discuss strategic evaluation. During that meeting, Susan mentioned that strategic evaluation represents the systematic assessment of strategic effectiveness and competitive impact. In the Vairos framework, evaluation goes beyond simple performance measurement to examine whether strategic initiatives are creating intended value and competitive advantage.

### Turning Data into Strategic Intelligence

Susan clicked open a slide. "Here's where measurement meets evaluation. "Think of it like this: measurement tells us what is happening; evaluation tells us why it matters."

Jim raised an eyebrow. "So, measurement's about tracking the numbers?"

"Yes," Susan replied. "It's all about quantifying performance, whether it's revenue, customer satisfaction, or operational efficiency, against our goals and benchmarks."

Megan jumped in, "But evaluation adds the context."

"Right," Susan said. "Without evaluation, data is just data. Evaluation adds the interpretation, it connects those numbers to strategy, helps us understand competitive impact, and shows if we're really progressing toward our mission."

Jim leaned in. "So how do we evaluate beyond just the numbers?"

Susan smiled. "By looking at different dimensions. First, we assess strategic alignment. Are we aligned in our mission? Then, competitive impact; are we gaining market share or improving our position? Resource efficiency is next. Are our agreements giving us the best return? And we look at stakeholder value, are we creating value for customers, employees, and investors?"

Megan nodded. "It sounds like we're constantly learning from this process."

"Yes, that's right," Susan said. "That's organizational learning. We assess the knowledge and capabilities we're building as we execute strategy. We also need to watch how well we're adapting to market changes; that's environmental adaptation."

Jim looked impressed. "So, in the end, evaluation gives us the insight to adjust, not just track."

"Precisely," Susan replied. "Without evaluation, measurement is just a report card. With it, we get strategic intelligence that guides our next steps."

## Streamlining Strategy Evaluation

Susan clicked to the next slide. "Let's dive into evaluation now. It's more than just tracking performance; it's about understanding why things are happening and how to adapt."

Jim nodded. "So, a deeper dive into the data?"

Responding to what Jim had said, Susan asked "First, we do a performance analysis, reviewing metrics and trends. Then, we add context: how does this performance fit within the market landscape? We also gather stakeholder feedback from customers and employees and run a comparative analysis to see how we measure up against competitors."

Megan smiled. "So, you're looking at both the data and the bigger picture?" She had asked a noteworthy question, and to this, Susan replied, "Then we look at trends. Is performance improving or stagnating? And impact assessment: how are we affecting the market and organization?"

Jim grinned. "Got it. So, tracking overall performance. What about specific initiatives?"

"Next up: strategic initiative evaluation. We assess objective achievement, implementation quality, and resource utilization. Are we using resources effectively and sticking to the plan? And, of course, timeline performance, is everything on track?"

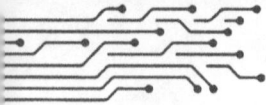

Megan added, "And the impact on stakeholders?"

"Definitely," Susan confirmed. "We also assess competitive response, how are competitors reacting to our moves?"

Jim raised an eyebrow. "But it's not just about progress, right? There's the financial side too?"

"Yes, there is," Susan said. "We look at ROI, cost-benefit analysis, market value impact, and revenue attribution, where is the money coming from? Finally, opportunity cost, what did we give up getting here?"

Jim leaned back. "It's all about understanding the impact and the cost."

Susan nodded. "And that's how we evaluate strategy: a comprehensive view of both performance and financials."

## Evaluating Strategy and Market Impact

Susan flipped to the next slide. "Let's talk about competitive and market evaluation. First, we look at market position, how are we shifting in terms of market share and competitive standing?"

Jim nodded and calmly said, "Are we gaining ground on competitors?"

"Yes," Susan continued. "Then, customer value creation: how are our initiatives improving customer satisfaction? We also check the effectiveness of differentiation, are we standing out enough in the market?"

Megan raised her hand. "What about how the market reacts to us?"

"Great point," Susan replied. "We need to assess market response. How are customers and the market reacting to our moves? And finally, competitive advantage: Are we creating lasting advantages over our competitors?"

Jim leaned forward. "What about our own internal growth?"

"Right," Susan said. "That's where organizational capability evaluation comes in. We look at skill development, how are we enhancing our

team's capabilities? Also, process improvements: are our operations becoming more efficient?"

Megan added, "And the culture?"

"Yes," Susan replied. "We assess cultural impact and employee engagement. Are we fostering a positive environment? We also evaluate system enhancements: how are our systems and infrastructure evolving to support growth?"

Jim smiled. "And all of this needs to happen regularly?"

"Yep," Susan said. "We set quarterly reviews for performance tracking, and annual evaluations for a deeper dive into strategic effectiveness. After each project, we do post-implementation evaluations, and at key milestones, we reassess our strategy."

Megan noted, "And you track both hard numbers and softer factors?"

To this, Susan replied, "We use quantitative metrics for clear targets, qualitative indicators for strategic quality, and compare against competitive benchmarks. Plus, we always consider stakeholder expectations and historical performance. How have we improved over time?"

Jim looked impressed. "This really ties everything together."

Susan nodded. "It's about understanding both market impact and internal progress so that we can adjust and improve continuously."

## Enhancing Evaluation for Strategic Success

Susan brought up the next slide. "Now, let's look at common evaluation mistakes. First, metrics without context. You see, numbers are meaningless unless we understand their strategic impact."

Jim nodded. "So, it's not just about tracking data, it's about why it matters."

"Exactly," Susan continued. "Then, short-term bias, focusing on quick wins without considering long-term effects. We also have attribution errors, where we misattribute outcomes to the wrong initiatives."

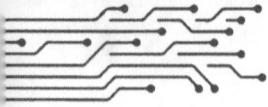

Megan added, "I see. And confirmation bias, seeking info that just confirms what we already believe?"

"Right," Susan said. "And stakeholder neglect, not factoring in the impact on all stakeholders, like customers or employees."

Jim raised his hand. "So, what's the key to successful evaluation?"

Susan smiled. "The success factors start with a systematic approach; structured evaluations give consistent insights. You also need a balanced perspective, considering multiple dimensions and viewpoints."

Megan nodded. "And being honest about both successes and failures."

"Yes, that's right," Susan said. "We also need a learning orientation, using insights to improve strategy and execution. And finally, action orientation, turning insights into real improvements."

Jim asked, "How does this tie into strategy?"

"Great question," Susan said. "Evaluation feeds into strategy refinement. It helps with future planning, resource reallocation, and the development of organizational capabilities. Most importantly, it's key for risk management, identifying and addressing risks before they become problems."

Megan smiled. "Sounds like evaluation isn't just a checkmark, it's a tool for improvement."

Susan nodded. "Exactly. It's how we get better with every strategic move."

## Feedback

Later that week, they started talking about feedback. Susan mentioned that it represents the critical connection between strategic execution and strategic learning, completing the Vairos planning cycle by transforming experience into insight. The latter helps improves future strategic planning and execution. In the Vairos framework, feedback is not closure but continuity, simultaneously ending one planning

cycle and beginning the next.

## Feedback as the Engine of Strategic Learning

Susan turned the page. "Here's something most organizations overlook; feedback. It's not just the end of a process; it's what keeps strategy alive."

Jim nodded. "So, it's about reviewing what worked and what didn't?"

"Yes, that's right," Susan said. "Feedback provides closure and opening, we wrap up one strategic cycle while learning for the next. It drives learning creation, turning execution into lasting organizational knowledge. It's also about performance improvement, spotting what can be done better, and cultural development, reinforcing a mindset of continuous growth. Over time, it builds institutional memory, so learning doesn't disappear when people move on."

Megan leaned in. "Where does all this feedback come from?"

"Everywhere," Susan replied. "From team retrospectives, leadership debriefs, and automated dashboards, to stakeholder input and competitive intelligence. We even use cultural surveys to see how strategy feels from the inside."

Jim smiled. "So, teams actually sit down and talk through results?"

"They do," Susan said. "Every quarter, teams hold retrospectives, 60 to 90 minutes of honest, data-backed reflection. They ask: What did we achieve? What fell short? What surprised us? And what did we learn? The key is documentation, or capturing insights systematically so they inform the next cycle."

Megan nodded. "And those conversations turn into actions, not just notes?"

"Exactly," Susan said. "The goal isn't to admire the data, it's to turn reflection into improvement. Feedback is how strategy becomes smarter every time we execute it."

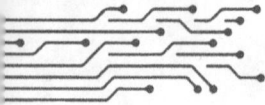

## Leadership Feedback as the Pulse of Strategy

Susan advanced to the next slide. "Feedback isn't just for teams, it's essential at the leadership level too. It keeps decision-making sharp and strategy aligned."

Jim crossed his arms. "So, we're talking about holding ourselves accountable?"

"Yes, that's right," Susan said. "We start with strategic decision reviews, a structured look at how well our choices played out. We ask: Did our strategies deliver the promised results? Were our assumptions, right? Did we invest in the right priorities? And what external shifts changed our course?"

Megan smiled. "Sounds like a reality check."

"It is," Susan replied. "We call them quarterly boardroom check-ups. The focus is on truth over defense; no excuses, just clarity. That's how we promote decision quality improvement, learning to make smarter calls each time."

Jim glanced at the data visualization on screen. "And I'm guessing the dashboards play a role?"

"Absolutely," Susan said. "Automated feedback systems give us constant visibility. Dashboard analytics show trends, KPIs track effectiveness, and exception reporting flags problems early. Then we use trend analysis to spot patterns and build a narrative around what the data means."

Megan leaned forward. "So, feedback doesn't stop inside the boardroom?"

"Not at all," Susan said. "We integrate stakeholder feedback, the customer voice, the employee perspective, partner input, even the community and investor viewpoints. Together, they give a 360-degree view of how strategy lands in the real world." Jim nodded. "And the cultural side?"

"That's huge," Susan said. "We run engagement surveys and culture assessments to see how strategy affects values and morale. We also gather input on communication effectiveness and change management, and how well people understand and adapt to

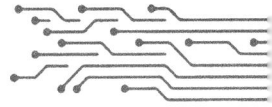

direction. And finally, we track learning capability, meaning how fast the organization grows from experience."

Megan smiled. "So, feedback becomes the organization's pulse; constant, honest, and forward-looking."

"You got it," Susan said. "When leaders embrace feedback, strategy stops being a plan on paper; it becomes a living, learning system."

Turning Feedback into Strategic Intelligence

Susan stood at the whiteboard, circling a phrase: "Feedback without action is noise."

"Collecting feedback is easy," she began. "Processing and applying it, that's where the real strategic value lies."

Jim nodded. "So, what does that process look like?"

"First, we start with pattern identification, looking for recurring themes across all feedback sources," Susan explained. "Then we move to insight synthesis, turning scattered observations into coherent lessons. From there, it's action planning, translating what we've learned into tangible improvements. We capture knowledge systematically, so it feeds into future planning, and we strengthen organizational capability by making feedback a learned skill."

Megan leaned forward. "So, feedback actually shapes strategy?"

"Exactly," Susan replied. "We integrate feedback into every planning cycle. It sharpens assessment, improves decision quality, refines execution, evolves our measurement systems, and, most importantly, shapes culture. When teams see feedback leading to visible change, they start to value it."

Jim pointed to the word learning on the screen. "That sounds like the foundation of a learning organization."

Susan smiled. "It is. Learning organizations build feedback habits into everyday routines, create psychological safety for honest input, and focus on learning, not blame. They institutionalize continuous improvement and ensure insights are captured through strong knowledge management systems."

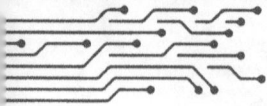

Megan looked thoughtful. "And what gets in the way?"

Susan clicked to the final slide. "Common feedback failures," she said. "Things like defensive responses, where honesty is punished. Information overload, where there's too much data but no synthesis. Attribution errors occur when we draw the wrong conclusions. Action paralysis, where insights go nowhere. And worst of all, memory loss, when valuable lessons are never documented."

Jim exhaled. "So, the goal isn't just collecting feedback, it's turning it into intelligence that shapes the future."

"Yes, that's right," Susan said. "When feedback becomes a system, rather than a ceremony, strategy becomes self-correcting. That, in a nutshell, is how organizations stay adaptive and alive."

## Feedback as the Engine of Strategic Renewal

In the final strategy session, Susan drew a loop on the board and labeled each stage:

Assess → Decide → Execute → Measure → Evaluate → Feedback → Assess.

"This," she said, "is our strategic flywheel. Feedback isn't the end; it's the bridge that turns experience into improvement."

Jim nodded. "So how do we make it work in practice?"

"Through systematic processes," Susan replied. "Regular, structured reviews ensure that feedback is collected, processed, and applied, instead of merely discussed."

Megan added, "That requires culture, too; people have to feel safe when giving honest input."

"Yes, that's right," said Susan. "Cultural support makes feedback real. But it only matters if it drives action, turning insight into tangible improvements. Leaders must model feedback use, showing that learning is valued more than perfection. And we integrate that knowledge directly into our systems, so each cycle strengthens the next."

Jim glanced at the diagram. "So, every turn of the flywheel builds

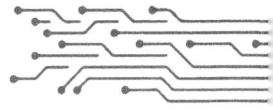

capability?"

"Yes," Susan said. "Effective feedback transforms strategy from an event into a continuous capability. It ensures every success and failure feeds the next plan, turning experience into advantage and keeping the organization adaptive, learning, and alive."

## Organizational Strategy

In one of their last discussions, Jim, Susan, and Megan talked about organizational strategy. Susan mentioned that this encompasses the fundamental approaches that organizations use to create competitive advantage and achieve long-term success. In the Vairos framework, organizational strategy provides the overarching guidance for all strategic decisions and initiatives.

### Balancing Evolution and Revolution in Strategy

Susan brought up a slide showing two diverging arrows. "Every organization faces this choice," she began. "Do we evolve what we have, or revolutionize who we are?"

Jim leaned forward. "Evolution sounds safer."

"It is," Susan said. "Evolutionary strategies focus on steady improvement, refining capabilities, and strengthening what already works. But sometimes," she paused, "the market shifts faster than we can adapt."

"That's when we need a revolutionary strategy," Megan added. "A bold move that redefines our position and capabilities."

"Yes, that's right," Susan replied. "Choosing between them, or blending both, depends on market conditions, organizational readiness, and competitive position. We also weigh stakeholder expectations and environmental pressures like regulation or disruption."

Jim studied the following diagram. "So, it's not either-or?"

"No," said Susan. "The best organizations do both. They evolve the core while revolutionizing the parts of the organization in need of dramatic improvements. Stated another way, they drive continuous

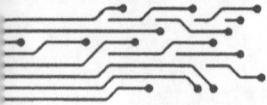

improvement in today's business and look for transformative projects for tomorrow. Through smart portfolio management, balanced resource allocation, and targeted risk control, they stay stable enough to compete, and bold enough to lead."

Susan gestured toward the next slide. "At Vairos, organizational strategy isn't just direction, it's the framework that guides every planning activity."

Megan nodded. "So, everything flows from that choice between evolution and revolution?"

To this, Susan responded, "Our assessment identifies where we can optimize what works and where we need true transformation. Then, our decision criteria ensures every choice supports the organization's larger objectives."

Jim added, "And once we decide, execution follows that same logic?"

"Right," Susan replied. "Execution planning aligns with the chosen strategic path, steady improvement or bold change. Measurement systems track the right kind of progress for each. And through learning integration, we capture insights from both approaches, strengthening our ability to optimize and transform simultaneously."

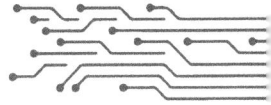

# Evolutionary Strategies

Then, they moved on to discuss evolutionary strategies focused on systematic improvement and optimization of existing organizational capabilities and market positions. Megan started off the conversation by stating that these strategies emphasize measured progression, risk management, and sustainable competitive advantage through excellence in current operations.

## Choosing and Executing Evolutionary Strategies

Susan clicked a new slide. "Not every strategy needs to be disruptive," she began. "Evolutionary strategies work best when the environment is stable and performance is solid, but improvement is still needed."

Jim leaned forward. "So, steady growth rather than reinvention?"

"Yes, that's right," Susan said. "They're ideal when markets are predictable, when we're consolidating after a big change, or when our stakeholders prefer measured progress. Think optimization, not overhaul."

Megan pointed at the chart. "What kinds of strategies fit that model?"

"Several," Susan explained. "Operational excellence, customer experience optimization, geographic expansion, talent development, digital maturity, and process innovation. Each focus on refining what already works."

Jim smiled. "And implementation?"

"Simple but disciplined," Susan replied. "We drive efficiency improvement, quality enhancement, cost optimization, and smart automation integration, all reinforced by performance management systems. Evolutionary strategy is about mastering the fundamentals, so growth becomes repeatable and sustainable."

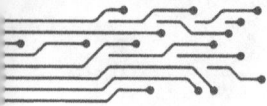

## Evolving Through Strength - Customer, Capability, and Digital Growth

Susan continued addressing the slide. "Evolution isn't just about tightening operations," she said. "It's also about deepening relationships and strengthening foundations."

Megan nodded. "So, that means customer experience comes first?"

"Right," Susan replied. "We focus on service quality, relationship development, and value enhancement, making every customer interaction better without reinventing the business. Add channel optimization and loyalty programs, and we turn satisfaction into long-term retention."

Jim pointed to the next section. "And growth?"

"That's where geographic and market expansion comes in," Susan said. "We grow through adjacent market entry, regional expansion, and partnership development, all by leveraging what we already do well."

Megan smiled. "And inside the organization?"

"Capability development keeps it sustainable," Susan continued. "We build leadership pipelines, enhance skills, and modernize technology, investing in digital integration, data analytics, and cybersecurity. The result is progress that compounds."

Jim leaned back. "So evolutionary strategy is steady but powerful."

Hearing this, Susan said, "It manages risk, uses resources wisely, builds confidence, and strengthens what competitors can't easily copy. These include our people, systems, and customer relationships."

## The Discipline of Steady Progress

Jim leaned forward as Susan spoke. "Evolutionary strategies look simple," she began, "but they demand discipline, the kind that keeps momentum long after the excitement fades."

Megan nodded. "So, it's not just about improvement projects?"

"Exactly," Susan said. "It's about consistent investment, clear measurement systems, and careful change management. You must track incremental progress, monitor competitors, and stay patient, because the results take time."

Jim smirked. "And I'm guessing there are plenty of traps."

Susan clicked to the next slide. "Five big ones: the complacency trap, incremental thinking, underinvestment, competitive blindness, and change fatigue. Each can stall momentum before real progress takes hold."

Megan leaned in. "So how do you stay on track?"

"With leadership commitment and cultural alignment," Susan replied. "You build systems for improvement, measure relentlessly, and keep stakeholders engaged. Evolutionary success isn't flashy; it's earned through persistence and precision."

Jim smiled. "Slow, steady, and unstoppable."

## Revolutionary Strategies

One of the last few things discussed was revolutionary strategies that transform organizational capabilities, business models, or competitive positioning. These strategies are designed to create breakthrough competitive advantages, respond to existential threats, or capitalize on significant market disruptions.

### Breaking the Mold - When Evolution Isn't Enough

Susan paused before moving onto the next slide. "Sometimes," she said, "evolution just can't keep up. That's when revolutionary strategy becomes the only path forward."

Jim raised an eyebrow. "You mean when the old playbook stops working?"

"Yes, precisely that," Susan replied. "When markets disrupt, technology transforms, or customers evolve, small improvements won't cut it. You need to rethink the game, not just play it better."

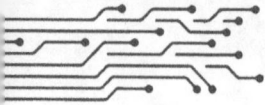

Megan looked intrigued. "So, what does that look like in practice?"

Susan smiled. "It could mean launching new ventures, transforming the business model, or adopting disruptive technology. Sometimes it's strategic M&A, a full brand reinvention, or even creating an entirely new market."

Jim crossed his arms. "Sounds risky."

"It is," Susan admitted. "But when you've hit a growth ceiling or face an existential threat, risk becomes a necessity. That's why smart organizations set up innovation labs, run market experiments, and form strategic partnerships before the pressure peaks."

Megan nodded slowly. "So revolutionary strategy isn't chaos, it's bold, structured reinvention."

After this, Susan clicked to the next slide and said, "It's about changing the rules, before the market changes them for you."

## Redefining the Rules - Strategic Revolution in Action

Susan swiped to the next slide. "Revolutionary strategies aren't just about change; they're about reinvention across everything the organization does."

Jim leaned forward. "Everything? That sounds massive."

"It can be," Susan agreed. "Take business model innovation for instance. We're talking about redesigning the value proposition, transforming the revenue model, and even restructuring operations. And it's not done in isolation: partnerships and platforms often redefine the entire industry value chain."

Megan interjected. "So, it's not just internal. Technology drives this, too?"

To this, Susan replied, "AI, automation, digital platforms, data monetization, and even blockchain can create entirely new ways to deliver and capture value. These aren't tweaks, they're leaps."

Jim nodded slowly. "And you coordinate this across the organization?"

"Yes," Susan replied. "Portfolio transformation is critical. That includes acquisitions, divestitures, joint ventures, strategic partnerships, and the development of ecosystems. Each move unlocks new capabilities and markets."

Megan raised an eyebrow. "And the brand?"

Susan smiled. "The brand and identity often need revolution too: redefining mission, repositioning in the market, shifting culture, reorienting stakeholders, and even rethinking communication strategies. Every layer, from operations to identity, can evolve simultaneously."

Jim chuckled. "So evolutionary strategies tune the engine, but revolutionary strategies replace the engine entirely."

"Exactly," Susan said, clicking to the final slide. "It's about changing the rules of the game before the market forces you to."

## Making Big Changes

Susan flipped to the next slide. "Revolutionary strategies are exciting, but risk management is crucial. You can't just leap blindly."

Jim frowned. "So how do we manage the risk of failure?"

Susan explained, "We start with staged implementation, testing concepts before committing major resources, and a portfolio approach, pursuing multiple initiatives at once. Every experiment teaches us something, and if one fails, we quickly reallocate resources. When something works, we scale it fast."

Megan leaned in. "And how do we actually make it happen?"

"Clear vision communication is key," Susan said. "Change management helps the organization adapt, resources are focused on revolutionary projects, and talent development ensures we have the skills to succeed. And remember, metrics must match the scale of transformation, not just incremental improvements."

Jim shook his head. "I can see how this could easily go wrong."

"Absolutely," Susan replied. "Common pitfalls include under-investing,

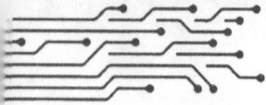

over-controlling, poor timing, ignoring culture, and failing to connect new initiatives to existing operations. That's why leadership, courage, and cultural flexibility are critical, along with smart timing and strong execution."

Megan nodded thoughtfully. "So, we don't abandon evolution entirely?"

"Not at all," Susan said, pointing to a diagram. "A dual-track strategy balances optimizing what we do now while developing revolutionary capabilities. Resources, culture, and timelines must all be coordinated, and lessons from incremental improvements feed the big changes. That's how we innovate without losing our footing."

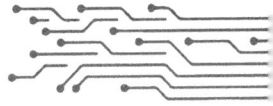

# Conclusion

To sum it all up, the Vairos Planning System provides a comprehensive framework for data-driven strategic planning that bridges the gap between organizational capability and market opportunity. By systematically moving through the phases of Assess, Decide, and Execute, organizations can transform data into a strategic advantage while building the capabilities necessary for sustained competitive success.

The power of Vairos lies not in any single component but in the integration of analytical rigor with execution discipline. Whether pursuing evolutionary improvement or revolutionary transformation, the framework provides the structure and tools necessary for strategic excellence in dynamic competitive environments.

Strategic planning is not a destination but a capability, the ability to systematically assess reality, make informed decisions, and execute with discipline while learning and adapting continuously. The Vairos framework builds this capability by providing transparent methodologies, practical tools, and integrated processes that transform strategic planning from an episodic activity into a continuous source of organizational advantage.

Victory comes to those who plan systematically, decide courageously, and execute with discipline. The Vairos Planning System provides the methodology for achieving victory in the right and opportune moment.